Biomimetics of Motion

Sandra Persiani

Biomimetics of Motion

Nature-Inspired Parameters and Schemes for Kinetic Design

Springer

Sandra Persiani
Faculty of Architecture
Technical University of Munich
Munich, Germany

ISBN 978-3-030-06581-2 ISBN 978-3-319-93079-4 (eBook)
https://doi.org/10.1007/978-3-319-93079-4

Softcover re-print of the Hardcover 1st edition 2019

Printed on acid-free paper

This Springer imprint is published by the registered company Springer International Publishing AG part of Springer Nature
The registered company address is: Gewerbestrasse 11, 6330 Cham, Switzerland

Preface

Teaching is much—Excitement is everything
Johann Wolfgang von Goethe

Designers and engineers constantly strive for optimization, improving man-made artifacts in their functionality and shape. Whether in the initial or the final stages of creation, designing buildings or micromechanical devices, the architects of things choose between combinations of endless possible solutions. This process of selection is intuitively carried out, relying on experience and a 'mental archive' of observed combinations. When inspiration fails, we learn to fall back on pretested solutions: this is where the advocates of biomimetics turn to Nature. The living systems we have before us are in fact fine-tuned combinations of design features, that evolution has tested and adapted over millions of years of patient design.

This study focuses on the features of animals and plants that enable motion, relying on the scientific work of zoologists and botanists. The aim is to identify patterns in the combinations of kinetic features, which could be a great asset for any engineer, architect, designer, or artist who wishes to integrate motion in their work. The book can be read sequentially or can be used for inspiration as a catalog of features.

Writing on this topic while not being a trained biologist or zoologist has had its advantages and disadvantages. Identifying recurring patterns in a reality that is as interesting as it is complex has been a tiresome work raising many doubts along the way. Classifications are always a very individual choice of what to highlight and most of all what to exclude. Moreover, the field of biomimicry is unfortunately limited to the state-of-the-art biology relying on specialists' selection and understanding of the case studies. On many occasions, the studies that have effectively been carried out can appear to be few in comparison to the overwhelming variation and abundance of solutions evolved by Mother Nature. Few systems have been studied extensively, and often by the same authors. This analysis is therefore knowingly a partial point of view of a very multifaceted reality, aiming to highlight recurring trends and successful combinations of motion parameters in living systems. Although the contents will appear oversimplified to the experts, I strongly believe that an architect's perspective on nature in motion can still be of inspiration to many neighboring fields of study. I hope that this book will be useful to my fellow colleagues and, for those who have never researched biomimicry, to uncover a strikingly stimulating field to explore.

This book is based on part my Ph.D. thesis in Environmental Design, developed at the University of Rome La Sapienza in collaboration with the Technical University of Munich. The thesis entitled 'Autoreactive Architectural Components, Theories and Schemes for the Implementation of Kinetic Reaction with Zero Energy' was defended in Rome on July 11, 2016.

This work would not have been possible without the direct and indirect support of a great number of people and institutions, among which:

Professor Alessandra Battisti at the University of Rome La Sapienza, who has been for the last twelve years an unshakable reference as mentor;

Professor Tina Wolf, Prof. Tillmann Klein, and the rest of the team at the TUM Associate Professorship of Architectural Design and Building Envelopes for the many interesting discussions and good advice;

Professor Dr. rer. nat. Werner Nachtigall's work on bionics and biomimics, and more specifically his book *Biologisches Design, Systematischer Katalog für bionisches Gestalten*, which has been a great source of inspiration;

Pierpaolo Riva, my valuable and patient editor who understood from day one what I have been trying to formulate for four years.

A special thanks to Luigi, Lisette, and Angelo. No journey is too long with the right company.

Munich, Germany

Sandra Persiani

Contents

Chapter 1
Introduction

Abstract Nature is a master of design, and has already solved many of the problems we are facing. However, how can we, as designers of man-made technology, identify which are the good solutions to mimic without becoming specialists? This work is an inspirational catalogue of kinetic features in nature aimed to assist engineers, architects, designers and artists to integrate motion in their work.

The adaptability of *Homo sapiens* and its development of technical and scientific knowledge has, under the last millenniums, undoubtedly brought it to be the most influential species on Earth. We have today, as far as we know, gained unprecedented advantages compared to all other species, to potentially secure our future in the evolutionary tree. We are the only species that actively achieves to manipulate and directly influence all factors responsible for natural selection:

- Resource availability. Humans are technically able to program and produce wished amounts of food, are aware of the state and quantity of the resources available on the planet, can strictly control their own habitat and sexual reproduction;
- Environmental conditions. Although humans have no direct control over the weather conditions, the technology and combined knowledge allows forecasting and preparing for upcoming changes. Moreover the artificial environmental conditions created through cities, buildings and clothing allow further adaptation, allowing this species to be present in all climatic conditions across the globe.
- Biological factors. *H. sapiens* is today at the top of the food chain in all ecosystems and can actively manipulate and control an important part of the most threatening diseases.

Still, although our species possesses such extraordinary advantages, its survival is today at stake. The technological and scientific advancement has gradually brought us to abstract ourselves from the natural and physical conditions imposed on us by nature recreating our own artificial environment to please our own needs. During our recent evolutionary history, we have thought we could outperform nature and

S. Persiani, *Biomimetics of Motion*,
https://doi.org/10.1007/978-3-319-93079-4_1

that the world was exclusively here for our use. In reality, we have forgotten that we are still bound to the ecological laws: we are neither immune to natural selection, nor (for now) the fittest survivors on the long run. The real survivors on the Planet are not the most dangerous ones at the top of the food scale, but are organisms which have found the way to live in generations over millions of years without consuming their ecological capital (Benyus 2002): 'more and more' works but lasts not, while 'better and better' always works (Nachtigall 2000).

1 The Importance of an Interdisciplinary Approach

Defining is about clarifying borders, making them distinct and precise, sorting out grey areas, to achieve simplification and identify the all-powerful generalized principles. While the modern science uses concepts as 'categorization' and 'generalization', as principles that have been found to hold true in every special case, as building blocks to proceed in a logical analysis; literature on the other hand, uses 'generalization' with a negative connotation, meaning a process tending to cover too much territory too thinly to be convincing. It can be in fact argued that specialization tends to shut off the wide-band tuning searches, precluding further discovery of these generalized principles. As pointed out by Fuller and Applewhite (1982), we assume that the existing trend of narrowing down our specialization to specific fields of knowledge is a logical and desirable way to proceed, when in fact Evolution shows that overspecialization is what causes biological extinction as the concentration of only selected genes sacrifices general adaptability.

In evolution, generalization of features has been an advantage more than a flaw. In fact, some theories point to these undefined features as man's foremost evolutionary feature. We are not the swiftest runners, swimmers, climbers, have no specialized environment—but we are the only animal who can do all things if only adequately—and have thrived in all environments on the globe (Williams 2013). Even our hands, taken as the distinctive feature of our species, are the tools of a generalist: in fact, manual skill depends upon initiation and coordination of the cerebral system and not on the specialization of the hand itself (Cohen 2010).

In most technical contexts, an important share of the innovation lies in the creative and conceptual stages of the process and progress is most of the time brought forward through single design solutions. Introducing spin-off alternatives from other disciplines is therefore a common strategy to boost inspiration and innovation (Gruber 2011): keeping an open attitude towards different approaches and ways of thinking is of crucial importance. These disciplines, at crossroads between artistic, scientific and technological realities, are characterized by the search for general scientific rules on one hand, and by the difficulty, or impossibility, of finding universally true practices on the other.

With this mindset, this work pursues an interdisciplinary approach, striving to find rules recurring across different fields of interest, bridging theories of Evolution and design, biomechanics and design of motion in technology. The system of ordered patterns is purposely left open enough to accommodate the emergence of alternative ideas, in a synergetic vision of the designing and the research process.

2 Nature as a Model

We have since the industrial revolution fancied we have grasped nature's secrets being able to outrun nature playing with the same physical rules. Our machines move faster than any animal, our planes fly higher, our bridges cover greater distances. However, comparing proportionally and not dimensionally our inventions to nature's, technology appears as quite obsolete in comparison to the wonders nature has evolved. Spider silk is ten times tougher than high-performance artificial fibers as Kevlar (Agnarsson et al. 2010). Grass stems are proportionally more light, elongated and flexible than any bridge or tower built by man.

Biological organisms are living examples of an enormous variety of systems in an almost perfect energetic balance, which constantly and sophisticatedly regulate and adapt themselves to a multifunctional energetic input and output. As Evolution favors energetically fit organisms over more obsolete systems, energy harvesting is a crucial factor for optimizing the organism's overall capacities: no extra energy is wasted on functions that can be autonomously regulated. In fact there is no better design partner than nature. One major difference between nature and man-made technology is that nature rewards mistakes and learns from them (Fox and Kemp 2009).

2.1 *From Bionics to Biomimics*

The difference between Bionics and Biomimicry is slight but significant.

The initial concept of *bionics was an imitation of nature* with the aim to develop technology in such a way to equal nature's technology. The aim was to investigate functional skills and technical solutions in nature, seeking analogies with the human technical world, in order to identify general design principles in the view of a solution transfer of these features to create new technological concepts by a process of abstraction and selection (Gosztonyi et al. 2013).

A clear example are the many researches in the field of medicine and robotics aiming at reproducing as accurately as possible a moving organism, limb movements or facial expressions. In time, developing technology for the simple purpose of refining the imitation capacity lost significance and the concept slowly turned towards a solution-finding technique for solving specific technological issues.

Biomimics, or biomimicry, borrow ideas from nature without intervening in the ecological context, leaving it untouched and available for others to draw inspiration from and covers a very broad range of application fields. It is much more than looking at the shape of a flower and get inspired by it, aiming towards a deeper observation and understanding of nature, to learn from the fit solutions instead of searching what to extract from nature or harvest parts of organisms as raw material (*bio-utilisation*). Biomimics often proceed from biology to design, as a natural phenomenon suggests a new way of solving a challenge; but can also proceed from design to biology as a specific challenge in the technical world requires solution, which is searched for among various systems in nature.

Biomimicry can be applied at different levels depending on the accuracy of the imitation.

- Level 1: intuitive imitation of a natural shape. The Velcro© strips for instance mimic the hooks on the surface of burs.
- Level 2: imitation of a natural process. This is the case of most research in chemical contexts.
- Level 3: imitation of the working principles of ecosystems. This involves greater systems as economies, production cycles, information and transportation systems, etc.

2.2 The Pioneers of Bio-inspiration

The first pioneer of bionics we know of was Leonardo da Vinci. Fascinated by the flying capacities of birds, he was convinced that the human muscles would be strong enough to produce enough power to fly. Around year 1500 AD he wrote the first treaty in the field of bionics, the '*Codex on the Flight of Birds*', and based on the observations of bird's flight he drew several sketches of flying machines. Many were not usable, others worked mostly as an energy source (a steam engine) (Behling 2005).

In the 19th Century, almost 300 years after Leonardo, many personalities in the scientific world revived the interest in nature-inspired design.

Sir George Cayley, a pioneer of aeronautical engineering, found the way to understand aerodynamics and the concepts underlying the forces of flight, setting the bases for flying machines, followed by Jean-Marie le Bris who built a glider inspired by the lift effects exploited by albatrosses.

Engineer Carl Cullmann set the bases of building through force trajectories, after observing the distribution of material in the trabeculae of bones. His theories were later followed and implemented by engineers Maurice Koechlin in the designing of the Eiffel Tower and later by Pier Luigi Nervi.

The inspiration from the natural world is clearly visible in the architectures of the 19th Century architect and gardener Sir Joseph Paxton. The slender steel and glass structures of the Crystal Palace and his greenhouses refer directly to the textures and structures of plants and leaves. Antoni Gaudi built architectures inspired by organic natural forms and systems and pinpointed the direction of forces in the structures through a system of suspended chains.

Raoul Heinrich Francé, a 19th Century botanist and microbiologist, was in recent years rediscovered as the founder of modern Bionics (although the discipline officially starts in the sixties in the States). He noticed how similar issues in nature have related solutions and each process is repeated following its most functional forms, independently of the complexity of the organism (Francé 1920; Behling 2005). He was the first to recognize the importance of developing technically concepts and ideas borrowed from nature, turning the practice into a new field of scientific research.

In the 20th Century, Sir D'Arcy Wentworth Thompson, a pioneer of mathematical biology, explained scientifically the process of morphogenesis, describing the mathematical beauty of nature, inspiring generations of architects, engineers, mathematicians, artists and thinkers worldwide (Thompson 2006).

Richard Buckminster Fuller is among the first to have explained natural systems as working through economical principles of material and energy efficiency. He developed the concept of synergetic systems based on the observation of the environmental and universal systems surrounding us (Fuller and Applewhite 1982). *Synergetics* is the study of systems in transformation, how the energetic interaction between parts and with other systems affect the system as a whole in an unpredictable manner. The concept can easily be pictured as a recurring principle present in the world surrounding us, from the complex interrelation of meteorological factors predicting the formation of clouds, temperature fluctuations, air pressure in the atmosphere to the upcoming of species and modification of ecosystems. Also, as an architect, he based the construction of his famous geodesic structures on the development of simple geometrical bodies as tetrahedrons, octahedrons and spherical package systems, inspired by radiolaries.

The German architect and engineer Frei Otto searched to grasp and explain the secrets of nature, developing unprecedented lightweight tensile membrane structures.

Werner Nachtigall is a biologist and pioneer in the field of modern bionics committed to explain the most recent discoveries in zoology, botany, biology and biomechanics in a technical language for the scientific interdisciplinary transfer from the biological to the technical world.

2.3 New Boundaries

In contemporary science, bio-inspiration is once again becoming the object of interest of many researchers in a broad spectrum of fields of study. After decennia of a process leading to overspecialization, interdisciplinary inspiration and study is now reconsidered a strength.

Noticeable outcomes of this new generation of biomimics are first and foremost the new products from the field of material sciences, where natural and artificial materials 'borrow' specific features from nature.

The Lotus-effect surfaces manage to remain free of contaminants without the use of detergents thanks to the imitation of the particular surface of lotus leaves at microscope level. The effect was first noticed in the biological system and successively transferred for use into the technical world (Gorb and Speck 2017).

Other case studies include mimicking high-performance proteins in spider silk for high-strength synthetic fibers, study of antifreeze proteins in ectothermic animals for anti-icing coatings in engineering, adhesive proteins of blue mussels byssus plaque for adhesive systems in underwater applications, proteins as molecular switches and biosensors for nanosensor technologies (Grunwald et al. 2009).

In the field of fluid dynamics and biomechanics, the observation of the presence of bumps on the humpback whales' flippers have lead to demonstrating the advantage of using aerofoils with a tubercle leading edges to prevent the foils from stalling, and to new wind turbine technologies (Hamilton 2008).

Contemporary neural computing studies use nature-inspired algorithms as accelerated particle swarm optimization, bat algorithm, krill herd algorithm, flower algorithm, etc. applied with different degrees of randomization, exploration, and exploitation in order to increase the diversity of solutions during the computational search process (Yang et al. 2014).

As can be seen from these examples, most of the contemporary science and technology turn to biomimicry mainly to enhance and perfect technology's performance and efficiency. These features are a necessary step in the development of our artefacts and indeed allow us to become more efficient and sustainable by reducing our consumption of energy and resources.

As Evolution has shown that long-term survival depends on adaptive capacity rather than specialization, which is instead used to develop transient features aimed at temporary adjustment (on an Evolutionary timescale) to specific conditions, how can we design for better adaptivity?

3 Biomimics for Adaptivity

3.1 Micro and Macro

Our perception of what is perceived as micro or macro depends upon a reference in relation to which we define it. Changing the reference, its context or scale leads towards a modified notion regarding what these terms and their link are. Something that was micro in one context may become macro when related to a different scale reference (Bulatovic et al. 2016).

From an engineering point of view, adaptivity in artefacts can be achieved on two different scales: on a micro- and on a macro-scale.

In the first case, *micro-scale flexibility* acts through the engineering of materials and of their behavior on a molecular and Nano scale. This field of expertise mainly pertains to chemistry, material engineering, biology and genetics.

In the second case, *macro-scale flexibility* is achieved through the design of a product and its parts. A great number of fields of expertise are touched by this topic, in particular all fields related to engineering, product design, architecture, mechanics, etc.

Adaptivity on a micro- and on macro-scale are however interrelated and complementary. In Nature's constructions, form and function are integrated in a same system, avoiding maximization of individual functions (which might harm other functions), rather optimizing the whole (Nachtigall 2000). There is an extremely refined combination of high-tech and low-tech solutions in laws of nature, complementing each other. Technology also resembles nature in this aspect: a building for instance is a sum of low- and high-tech solutions working together and resulting in better quality and efficiency than if only one type of component was used (Nachtigall 2003).

This work aims to address adaptivity on a macro-scale. In order to do so however, inspiration and research is conducted across scales as from a geometric and functional point of view, the differences between scales blur and become synergic (Bulatovic et al. 2016).

Adaptivity on a macro-scale, through design, can be achieved mainly in two ways:

- through disassembly;
- through motion.

The difference in adaptation through disassembly and through motion is primarily the time-factor. While flexibility through motion is a response to immediate pressures, flexibility through disassembly allows to prolong the use of a system or of its parts, and to adapt over a longer timeframe.

3.2 Adaptivity Through Disassembly

Although design for disassembly (DFD) was initially driven by economic aspects linked to the optimization of the production process, the field is today realizing the huge potential of these techniques to transform products' end of life and optimize reuse, remanufacturing and recycling of materials, components and sub-assemblies (Lowe and Bogue 2007). In a world with finite resources, growing population and consumption, this topic is becoming of critical importance as all our products are quickly being outpaced and exchanged for newer, better performing devices.

A product can be disassembled either by destroying the product and recover useful parts or materials, or by *reversing the assembly process* (Barsan et al. 2007). This second type of disassembly can be seen as an adaptive capacity through a process, rather than an adaptation of the organism itself.

Reversing the assembly operations can many times be a difficult task and requires most products to be designed for it. The benefits of efficient disassembly are however important allowing to consistently reduce waste, provide ready components for future re-assembly of other products, simplified recycling of unused parts, enhances maintainability, repair and substitution of obsolete parts (Barsan et al. 2007), and thus to potentially prolong the usage of the product.

Temporary architectures were historically used for being pragmatic: repaired and replaced, they gave better answers to functional problems than most of the present day buildings designed as timeless monuments (May 2010). Today, computer-aided manufacturing processes are revolutionizing the design and construction of the buildings and their parts allowing simple manufacturing and construction of complex shapes, changing dramatically the old paradigm of designing buildings as made of many equal parts (Knippers and Speck 2012). A number of experimental structured built by the Institute of Building Structures and Structural Design (ITKE) together with the Institute of Computational Design (ICD) of the Faculty of Architecture and Urban Planning at the University of Stuttgart have shown interesting projects of lightweight bio-inspired parametric structures which reinterpret the traditional carpentry joints. The 2010 research pavilion works as a radially interconnected coupled arch system of elastic bending thin plywood strips.

3.3 Adaptivity Through Motion

Kinetics is in the first place a way to introduce function and adaptability in our objects to keep up with the changing human patterns of interaction with the built environment (Zuk and Clark 1970). Most man-made artefacts are designed to move. If we look at the objects surrounding us, a great majority of our tools embed motion in their design in at least one of their variants: purses have openable zips and flaps, pens are twisted or pushed to uncover the ballpoint, paperclips bend, windows rotate on hinges as books do on their spine, etc.

Moving components accommodate specific ranges of activities by changing geometry or orientation. On one hand this gives the objects their function and usability, on the other it makes them subject to use and wear and are often the first parts to break. When motion is made impossible, the usability of the object declines and the article becomes waste, hence the need of *integrating adaptivity through disassembly together with adaptivity through motion*. These two concepts can together achieve change, agility and liveness together with a broad range of benefits as aesthetics, function, flexibility, energy use, reliability, user friendliness and greener life cycles in our artefacts.

4 Biomimetics for Non-specialists

The most famous discoveries, from Newton's apple to the invention of Velcro®, come with an anecdote on how the inventor stumbled by chance upon the right solution to their problem. The path to the greatest scientific breakthroughs is paved with random events and coincidences as much as with hard work.

In order to maximize the chances of these 'fortuitous events' and to provide inspiration to designers and engineers, a growing number of works and initiatives are today available, aiming to diffuse bio-related research findings to other fields of study though catalogues of examples and systems (Nachtigall 2005; Kaiser et al. 2012). The most recurrent method brought forward in biomimics is the case-by-case study: most bio-inspired devices can directly refer to one nature-based inspiration source. *In most cases, attention is focused on the solution itself, and rarely on the selection process leading to a successful solution finding*. One thing is to observe a solution in nature and to find a technical application for it, which is the case of a few success stories (as Velcro® or WhalePower technology); another is to have a defined objective and search for a fitting solution, which is the problem most engineers face on a daily basis.

How can designers and engineers without any knowledge in biomimics or biology find a shortcut to the solutions fitting a specific need? How can we assert that the chosen example, or combination of solutions are the most appropriate for the intended goal?

This work aims to gather and categorize different aspects describing motion in living systems in an inspirational catalogue of kinetic biomechanic features aimed to assist engineers, architects, designers and artists to integrate motion in their work. Geometric, mechanic and rhythmic parameters are described in depth through examples from animals and plants, and are contextualized within an integrated mapping of successful biomechanic combinations. The schemes aim to identify for the designer those aspects with a high probability of working together efficiently, opening up for new solutions and a method of evolutionary problem solving.

References

Agnarsson I, Kuntner M, Blackledge TA (2010) Bioprospecting finds the toughest biological material: extraordinary silk from a giant riverine orb spider. PLoS ONE 5(9):1–8

Barsan L, Barsan A, Bolos C (2007) Design for disassembly as an essential step in fulfilling the eco-design goals: reuse, repair, re-manufacture, recycle. JIDEG: J Indus Des Eng Graph 3 (3):22–24

Behling S (ed) (2005) Bionic skins III. Seminarbericht Sommersemester 2005, Universität Stuttgart Institut für Baukonstruktion

Benyus J (2002) Biomimicry: innovation inspired by nature. Harper Perennial, New York

Bulatovic K, Bunjak K, Persiani SGL (2016) Symbiotism and the synergy of scale. In: Proceedings from the international conference and exhibition—scale of design from micro to macro: from furniture to urban design, STRAND, Belgrade, 1–2 December 2016

Cohen YA (2010) Human adaptation: the biosocial background, 1st ed. (n.d.), Aldine Transaction

Francé RH (1920) Die Pflanze als Erfinder, Kosmos, Gesellschaft der Naturfreunde, Stuttgart. Accessed 28 April 2016 from https://archive.org/details/diepflanzealserf00fran

Fox MA, Kemp M (2009) Interactive architecture. Princeton Architectural Press, New York

Fuller RB, Applewhite EJ (1982) Synergetics, explorations in the geometry of thinking (n.d.). New York: Macmillian

Gorb S, Speck T (2017) Biological and biomimetic materials and surfaces. Beilstein J Nanotechnol 8(1):403–407

Gosztonyi S, Gruber P, Judex F, Brychta M, Richter S (2013) BioSkin–Forschungspotenziale für bionisch inspirierte energieeffiziente Fassadentechnologien. Berichte aus Energieund Umweltforschung 46/2013, Nachhaltig Wirtschaften, Bundesministerium für Verkehr, Innovation und Technologie

Gruber P (2011) Biomimetics in architecture, architecture of life and buildings. Springer-Verlag, Wien. ISBN 978-3-7091-0331-9

Grunwald I, Rischka K, Kast S, Scheibel T, Bargel H (2009) Mimicking biopolymers on a molecular scale: nano(bio)technology based on engineered proteins. Philos Trans Math Phys Eng Sci 367(1894):1727–1747. Accessed from http://www.jstor.org.ezlibproxy1.ntu.edu.sg/stable/40485549

Hamilton T (2008) Whale-inspired wind turbines, mimicking the bumps on humpback-whale fins could lead to more efficient wind turbines. MIT Technological Review. Sustainable Energy. Accessed 28 September 2017 from https://www.technologyreview.com/s/409710/whale-inspired-wind-turbines/

Kaiser MK, Hashemi Farzaneh H, Lindemann U (2012) An approach to support searching for biomimetic solutions based on system characteristics and its environmental interactions. In: International design conference—design 2012, Dubrovnik, Croatia, 21st—24th May 2012

Knippers J, Speck T (2012) Design and construction principles in nature and architecture. Bioinspiration & Biomimetics n.7

Lowe G, Bogue R (2007) Design for disassembly: a critical twenty-first century discipline. Assembly Autom 4:285

May J (2010) Handmade houses & other buildings, the world of vernacular architecture, 1st edn. Thames & Hudson, London

Nachtigall W (2000) Das Große Buch der Bionik, Neue Technologien nach dem Vorbild der Natur, Deutsche Verlags-Anstalt, Stuttgart/München

Nachtigall W (2003) Bau-Bionik, Natur – Analogien - Technik. Springer-Verlag, Berlin, Heidelberg, New York

Nachtigall W (2005) Biologisches Design Systematischer Katalog für bionisches Gestalten. Springer Verlag, Berlin

Thompson D'AW (2006) On growth and form. Cambridge University Press, 1961 (Trans: Albus A). Eichborn AG, Frankfurt am Main

Williams C (2013) Origins of form, the shape of natural and manmade things, 1st edn. Taylor Trade Publishing, Lanham

Yang X, Karamanoglu M, Ting T, Zhao Y (2014) Applications and analysis of bio-inspired eagle strategy for engineering optimization. Neural Comput Appl 25(2):411–420

Zuk W, Clark RH (1970) Kinetic architecture. Van Nostrand Reinhold, New York

Chapter 2
Evolution and Body Plans, as Nature Designs

Abstract Nature *optimizes* and adapts by intervening on two different scales of design: optimization of the process and accumulation of single optimized solutions.

- *Optimisation through process* is achieved by Evolution;
- *Optimisation through single solutions* is achieved by producing a vast variety of combinations of characters.

These can be represented by identifying recurring patterns of solutions. This chapter aims to discuss adaptive design principles and to build up a systematic framework for the further deepening of patterns of solutions in the next chapters.

After 3.8 billions of years of evolution, trial and error, nature has worked out many long-lasting examples of energy-efficient systems that are adapted to live on our planet. The specific design of an organism is the result of the successful coexistence of different parameters that have co-evolved given the conditions at the base. In general terms, this does not imply that Evolution has found the absolute best solution possible, but rather how to make the best out of the conditions at any given time of the evolutionary history (Dawkins 1996).

The development and improvement of man's technical skills has often been described as a continuation of evolution with other means. The human species has progressively modified everything in its environment, from the artificial urban cityscapes, to its diet by selecting, genetically modifying crops and domesticating other species. Technology is now pushed towards a further boundary: boost the human body's capacities and augment its abilities (Art Science Museum 2017).

Man has often turned to Nature in seek for inspiration and solutions. Biomimics see in Nature a real-life encyclopedia of pre-tested and optimized technical solutions, which can consistently raise the chances of finding good solutions, as a shortcut in the designing process. Research on neural networks inspires problem-solving in the context of data processing, while single capacities of animals and plants inspire new proprieties in nano-materials, as gecko-feet inspired adhesives (Menon et al. 2004) and lotus-leaf inspired water repelling surfaces (Latthe et al. 2014).

S. Persiani, *Biomimetics of Motion*,
https://doi.org/10.1007/978-3-319-93079-4_2

Nature is in fact a highly creative concept innovation pool, but should not be confused with a collection of blueprint solutions as the ideas can usually not be directly transferred from one field to another. Solutions from nature can be translated to technology, provided that these are adequately grasped (Nachtigall 2003). While evolution copes with sustainability on very long terms, prioritizing survival rather than perfection, human technology strives for perfection at a given point in time, depending on short-time factors as local economies, availability of resources, cultural attitude to different issues and not least specific business objectives.

Seen from this perspective, one can reasonably argue that architects and engineers, who generally have absolutely no knowledge on biology, zoology and biomechanics, can be inspired by nature but rarely achieve a real biomimetic result. If the designing process starts with the study of a specific feature in a biological system, which is then translated into a technical device the biomimetic design has good chances to be successful. In most cases however, technologists are given a technical problem and must search for a bio-inspired solution. How can an architect or an engineer consciously choose which example to study and follow, and be sure that the chosen example is with good probabilities the most appropriate one to follow? To obtain the certainty of following a biomimetic process, a team either needs to integrate one or more specialists on the specific topic, or the technologists need to specialize further to get an insight into these matters. Basic research in Biology is hence a prerequisite for each biomimetic application, and Technical Biology (TB) and Bionics (B) are interdependent and complementary fields of study. One being the ground source and the other the vehicle of technological advancement.

1 Principles of Optimization

1.1 Cumulative Selection

In nature different kinds of selective processes achieve automatically generated non-random forms of order. Matter arranges according to its proprieties under the constant action of the physical forces—through *single-step selection*. Pebbles on a beach are automatically sorted according to their size and weight upon the action of wind and waves. Evolution uses another kind of sieving process, where the results of one sieving is the start for the next one—*cumulative selection*. Entities are sorted over many generations, rapidly cutting out from the probabilistic variation process incongruous combinations.

To explain the difference between single-step and cumulative selection, Dawkins (1996) takes up the probabilistic theory of a monkey writing the complete works of Shakespeare. Probabilistically, a monkey dashing through a typewriter and given infinite time could be able to write the complete works of Shakespeare. The chance of the scenery to occur is so tiny that even the age of the universe would probably be too low, but is technically not zero. This is a case of single-step selection: the monkey has to start the process over again each time until the correct combination is found. Now, if we limit the task of the monkey to a single sentence, not to the complete works of Shakespeare, and give him a typewriter with a restricted keyboard, the probability of achieving the correct combination rises abruptly. The time for a computer to write that same sentence as the monkey in a context of cumulative selection is about 11 seconds. The same computer, working at the same rate and forced to use the procedure of single-step-selection would need about a "million million million million million million years" to reach the same target.

1.2 Categorization and Information Processing

The more a definition is precise, allowing an easy categorization with no other alternatives—yes/no; black/white; positive/negative; inside/outside—the easier the processing of information. In fact, the emergence of the word *Type* is strongly connected to the reification of the world concomitant to the rise of scientific thought (Madrazo Agudin 1995). The concept of *Type* is widely diffused. It seems to arise in organisms with medium to high levels of intelligence as a way of processing information though categorization, recognition of recurring patterns and definition of connections between these patterns (Fuller and Applewhite 1982).

Nature itself uses recurring strategies as the multiplication of available basic elements to minimize the information without running into a consistent loss of data (Gruber 2011). A close translation of genetic coding to the technological realm can be found in the complex structures achieved by binary number systems. Computation gives us a sense of how complex macro-systems can be built with the use of very few basic building blocks. Zero-and one sequences create bits of information, grouping together to create bytes—representing orders in coding capable of describing everything from still to moving images, sounds, programs—creating more and more complex systems. According to Steven Vogel (Zinsmeister 2011), a fertile egg for the construction of a human contains around 1010 bits of information: an amazingly low amount of necessary information, compared to its apparent complexity. Which are the underlying building rules to achieve such control of complexity?

2 Principles of Differentiation

Evolution is based on the probabilistic trial system through which billions of random alternatives and combinations are tested over millions of years. Nature counts on small transcription errors as a way of experimenting, with a trial-error method, alternative solutions. This way, nature has been designing, molding and optimizing the complex elegance of the systems we observe today without any premeditated crafted design. The complexity of these structures is characterized by heterogeneity, which is unlikely to have arisen by chance alone because of the uniqueness of the arrangement. Of all the possible combinations, the existing ones are the only (or the few) which result in a specific proficiency (Dawkins 1996), as the exact arrangement of parts in a bumblebee that make it possible for it to fly.

Following concepts, derived from the field of genetics illustrate different ways of designing complexity with basic building blocks.

2.1 Differentiation on the Group Scale: Patterns and Categories

Information in natural organisms is processed though the categorization and recognition of recurring patterns and through the definition of connections between these patterns. The main strategies used to minimize information use structures based on the *multiplication of simple elements* among which:

- Symmetry. Radial and bilateral symmetry extend the possibilities of reducing the amount of information required (Gruber 2011).
- Simple rules binding elements into self-similar and repetitive structures, as fractal geometries, helixes of DNA, etc. (Mandelbrot 1983);
- Combinations of variations of simple elements to form dissimilar structures, as cellular structures organized to form different organs and body parts;
- Segmentation. Identical subunits used to form an organism or part of it, as the bones of a spinal cord;
- Multiple use of information. A single set of instructions, depending on how they are written may produce more than one structure, as monozygotic twins (possessing the same genetic material) which can develop to be quite different depending on the body's different answers to internal and external pressures.

2.2 Differentiation on the Individual Scale: Type

Cloning, a progressive *change in quantity*. A clone is an exact copy of a 'mother cell'. When a cell proliferates dividing itself into a number *n* of descendant cells becoming a developing organism, we can also say that the totality of these sub clones corresponds to the original 'mother-clone'.

Mutation, a *change in quality*. Mutation is the process through which an organism produces change. It can be seen as a small random "transcription error" during the duplication of a parental gene, creating a 'defective clone'.

Genes, intermittent *change in time*. The development stages of an organism are not homogenous in time. Different kinds of genes are responsible for controlling the development and have characteristic periods of action, being switched on and off according to what is needed. Their timing might be early, late, brief, prolonged, single-pulse, intermittent (etc.) and the result of their interaction is the development we can observe. All genes do not however have developmental functions (deciding body-plan, size, shape, constituent parts and so on): terminal target genes are switched on only in specific cells regulating very specific functions of the cells; while housekeeping genes regulate routine-functions as the organism's metabolism.

Probabilistically, the earlier a gene is switched on, the greater effect it has on the development on a system as an early gene would alter a larger amount of sub-clones than a later one. This is however not absolutely true as it depends on the action of the gene, which has a complex lifecycle more often than not: it might have a different proliferation rate, expression- and effect-zones might not coincide, cells may come under the influence of other morphogens and its expression might be intermittent in time.

2.3 *Frequency of Differentiation*

Naturally, the higher the frequency of mutations, the greater the possibilities for experimenting new combinations. In fact, the difference between natural mutations and laboratory mutations is merely one of frequency: the concentration of many mutations in a single organism in nature would probably be a handicap as the unknown consequences of the mutations might potentially endanger the organism. Instead, nature works probabilistically on populations of entities and over billions of years. Different theories have however been suggested concerning the nature of the mutations leading to evolution.

- Macro mutations: species have originated in single macro evolutionary steps; many late changes cannot accumulate to give an early one.
- Micro mutations: major morphological differences have been produced through the gradual accumulation of very small changes (neo-Darwinian view).

The distinction micro/macro mutation provides poor basis for progress as a classification of the mutations' 'effect magnitude' is impossible: small mutations might have big consequences and vice versa (Arthur 1997).

3 Principles of Selection

Mutation and selection are the two complementary aspects in Evolution, as in any other creative process as well. *Mutation* is the creating force, altering the developmental characters, creating morphological and substantial diversity. *Selection* is the destructive counterpart of mutation, regulating individuals and populations, testing to eliminate unfit combinations, and opening up new niches for other development explosions.

3.1 Selection Through Destruction

Selection, as for differentiation, can be found on many different levels, as well as internally or externally to an organism. In general terms, a selective agent is considered as any factor influencing or limiting the development, survival and reproduction of a given organism. *Internal* selection occurs when individuals with different genotypes at a given locus differ in fitness because they differ in their degree of internal coordination. If we think of organisms as machines, fitter genotypes correspond to machines whose parts integrate better with each other (Arthur 1997). *External* selection includes available energy sources, surrounding environmental conditions, hazards, competing organisms etc.

3.2 Selection Through Conservation of Tested Working Patterns

Selection is however not an altogether-destructive method of design, but can be seen more as a sieving of functional solutions out of the thousands of possible combinations. Nature is reasonably economical, not wasting time and energy searching for alternatives to old over-tested patterns. The elder the features are, the more conservative they are. This is why recent mutations rarely have effect on body plans. Probabilistically, the earlier a mutation occurs, the bigger overall effect it has as the new mutated gene will be duplicated into downstream-cells, if conserved during selection. On the other hand, recently developed features as size, shape, color or reproductive maturity have greater possibilities to mutate. A mutation in the early genes simply risks to be too disadvantageous for the organism.

Evidence of this conservative design can be seen in animal evolution, which was more creative in its early stages. In fact, the 'morphological designs' of the ca. 35 existing animal phyla arose in a burst of creative evolutionary activity about 500 millions of years ago and have little changed since. In particular, creative evolution requires a number of coexisting factors which rarely align, as:

- Ecological opportunities, as new habitats or niches to invade;
- An easily modifiable basic shape with unspecialized lineages;
- Great supply of genetic and developmental variants.

4 Morphology and Type

An organism can be seen as an inverted cone of cells. The zygote (fertilized egg) develops into an embryo and successively into a complete organism and an adult organism by producing clone offsprings of the parent-cell, which successively specialize on various tasks controlled by their internal genetic information-switch. Cells change character, first defining the general features of the body and step-by-step developing more specific features (Menges and Hensel 2008).

Type is born when a certain number of variations is produced. All patterns consist of experience recalls and reside in memory, and a pretty long sequence of data is needed to picture any reliable scenario (Alexander et al. 1977). A concept can be explained, but the definition alone is partial and incomplete without the direct experience. Whoever has learned to know a different culture or tradition, is aware of the existence of specific expressions and thinking paths for specific groups of individuals. It is very difficult to translate expressions, which are linked to specific languages, as the single concept is linked to a multitude of concepts and to a culturally different way of thinking. The meaning can be explained and described, but a foreigner will not be able to categorize and understand the concept without experiencing it directly and in many different contexts.

Morphology is the study of forms. The German language distinguishes *Gestalt*, a structured form which refers to something already formed, and *Bildung*, the formation process where the structured form changes in an ongoing process. In nature, an organism never crystallizes with one peculiar shape but is an ongoing process of metamorphosis comprised in a definite range of possibilities.[1]

Morphogenesis is the process, which controls the spatial organization of cell groups to form specific organs, tissue and overall anatomy, and is closely related to the concept of *Type*. Morphogenesis is not a reversible process, however the concept of having a multitude of more or less identical organisms grouping and ungrouping in different spatial arrangements to form different polymorphic combinations is interesting in any design context (Wolpert et al. 2011).

Morphospace. For a parent organism to produce an offspring of its own size, the young must undergo a metamorphosis before becoming an adult. The concept of “adult organism” is however a rough approximation of reality, suggesting the stabilization of the organism’s morphology. In reality, an organism passes during its lifetime through a series of forms in space, which can be represented as departures from an initial trajectory to a final one, the cloud of points representing all existent morphologies in the phylogeny[2] being the morphospace. Among the parameters

---

[1]The differentiated morphology of living organisms is determined by the genome, the combination of forces internal to their cells (chemical activity), and of external environmental forces (gravity). The effect of these forces is different depending on the size of the organism (Menges and Hensel 2008).

[2]Phylogeny is the evolutionary history of a kind of organism (Merriam-Webster dictionary).

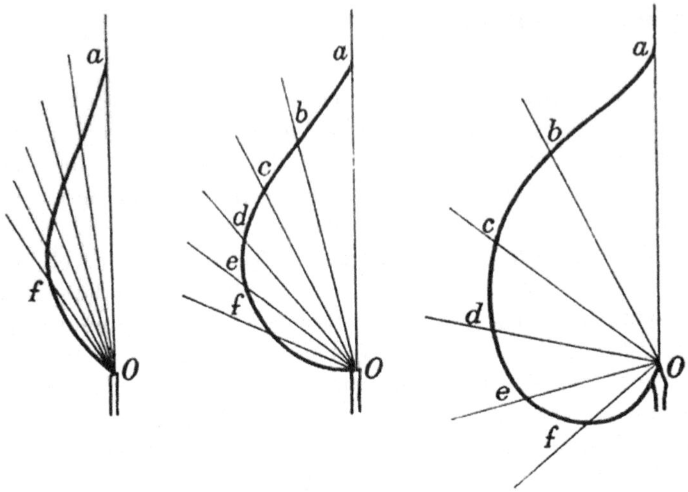

Fig. 1 Similar shapes which could represent the growth of a leaf, all belonging to the same morphospace (from Thompson 2006)

defining the morphospace are: the number of cells involved, type of cells, general form and symmetry, characters defining eventual segments and appendages (Arthur 1997) (Fig. 1).

Morphology and the design of primitives. Another way of describing the concept of Type is close to Locke's concept on the formation of Ideas: it results of the abstraction of the common features of a series of individuals, concentrating in one concept only what is common in all of them and distinguishing between the primary qualities on which the existence of the object relies: solidity, figure, number and motion; and secondary qualities which qualify the primary aspects: color, sound or taste (Madrazo Agudin 1995). In architecture and design, an example of this kind of thinking can be found in Greg Lynn's (2004) project for the Alessi Flatware: the architect imagines what he calls a "primitive design" containing all elements of its offsprings: 22 tableware (spoons, forks, knives, servers…) are designed by mutating the primitive into each final shape (Fig. 2).

Lynn (2004) describes the primitive as a virtual undefined entity, which is recognized by everybody and awaits to be unfolded and acquire its details. This concept very much reminds of the distinction between Body Plans and Characters in evolutionary biology.

Fig. 2 Flatware variation based on a common primitive (from Lynn 2008)

5 Body Plans, Functional Patterns of Combined Features

Classifications of natural systems can easily prove too generic to allow a deeper comprehension of the underlying schemes, or too rigid compared to the complex multiplicity of the biological elements. Some examples may perfectly fit into the individuated categories, though many more are probably on the edge between one or more definitions. In order to attempt a better understanding of those geometries, the following organization introduces two fundamental concepts distinguishing it from a common multi-aspect classification:

- Two homologous geometries distinguishing themselves by one or two different geometrical proprieties will further be called *sibling-geometries*. Supposing all basic geometrical shapes can be organized as linked to one another through their siblings, it should be possible to obtain any geometry by variating enough parameters (and therefore siblings) starting from any geometry, creating a circular- rather than a linear-family pattern;
- Two sibling-geometries are connected by a multitude of possible inter-geometrical shapes, which can be more or less close to one or to the other conformation.

Body plan refers to the organisms' possession of particular assemblages of homologous architectural and structural features (Valentine 2004). *Characters* are more recently developed features that adapt the Body Plans to new environments and niches. Body Plans from nature individuate unique combinations of the states of six major characters:

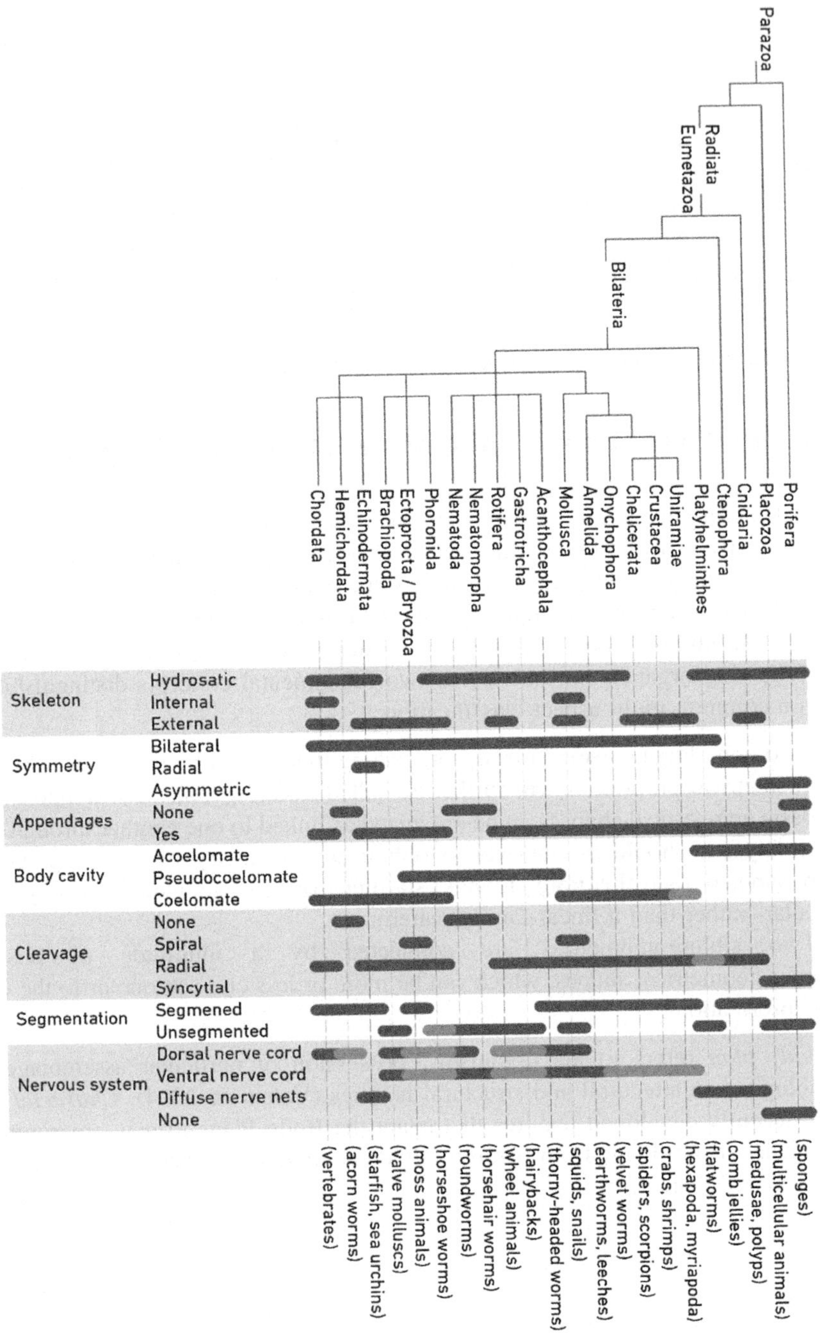

Fig. 3 Major phyla versus body plan characters (based on Arthur 1997)

1. Type of skeleton;
2. Symmetry;
3. Appendages;
4. Body cavity;
5. Cleavage pattern;
6. Segmentation.

All other features are acquired later and are evolutionarily transient: they originate, transform and may disappear again. In contrast, Body Plans have a high degree of conservation extending over hundreds of millions of years and across thousands of species (Arthur 1997). The body plan is established within the first stages of growth, a couple of months after fertilization. The rest of the period regards a change in complexity and proportions with an allometric growth of pre-formed parts. Special features adapt the general draft-plan to the environment: appendages appear first, then specialize becoming arms with hands/legs with hooves/wings with feathers/fins etc. (Fig. 3).

Parallels can be drawn between Evolutionary Biology and the history of man-made artifacts, from the genesis of typologies and further to different specifications and sub-categories. Type is, as a Body Plan, an adaptation to the exteriors and interior factors acting on an organism. Pressures, as evolutionary stresses, can be permanent or transient, physical, functional, social, economic, cultural, etc. influencing the design of our creations (Zuk and Clark 1970) (Table 1).

6 Energy, the Driver of Optimization

All selective agents, whether internal or external to an organism or a system, can be broken down to a common factor defining the energy-efficiency of the system and its parts. In nature, energy waste means extinction. A system, which does not rely on sustainable principles, is sorted away by natural selection and evolution. Any task performed in an organism must therefore be as precise, as rapid and as little energy requiring as possible (Nachtigall 2000).

Biological systems, being highly organized irreversible structures, inspired Ilya Prigogine to develop a theory of non-linear thermodynamics in states far from equilibrium, trying to explain the spontaneous process of self-organization of disordered structures into ordered structures as homeostasis, morphogenesis, pattern formation, coordination in movement, etc. What he called *dissipative systems* are open systems in constant energy exchange with their surroundings: if driven far from equilibrium, new situations and systems can be formed, stable to perturbations and ordered in time and in space. In other words, these systems also live in symbiosis with their environment (Reichl et al. 2004).

The Prigogine theory introduces the concept of irreversible systems based on instability and statistically explained in terms of probability. These systems include radioactive decay, solar radiation, weather systems and the emergence and evolution of life. In these scenarios, time is seen as irreversible, playing a constructive

Table 1 Comparison of common building rules in nature and in technology (based on Nachtigall 2000, and Menges and Hensel 2008)

DESIGN OF NATURE	DESIGN OF TECHNOLOGY
Integrated construction It is difficult to distinguish the limits of single elements in a whole	**Additive construction** Single parts of different materials compose a working system
Multi-functionality One element has more tasks, each problem has a network of systems solving it Ex. Claws are used for prey, for defense, etc.	**Mono-functionality** Technical devices fulfil one or few tasks striving for precision and error-free manufacturing
Organism adapted to the Environment Solutions evolved in many variations to fit a specific environment Ex. Golden eagles' talons (*Aquila chrysaetos*) are hooked to grab rocks. The Osprey's talons (*Pandion haliaetus*) are long and hooked with a reversible toe to grab slippery fish	**Environment adapted to man** Ex. Contemporary air-conditioned structures **Man adapting to the environment** Ex. Naturally ventilated traditional buildings
Energy effectiveness It is the strongest evolutionary challenge	**Energy dissipation**
Use of renewable energy Direct through photosynthesis Indirect through feeding on plants and animals	**Use of fossil fuels, use of renewables**
Temporal Organisms decompose when not used	**Durable** Synthetic objects become waste after end of use
Total recycling No waste, any superfluous removed substance is at another organism's disposal to integrate	**Waste accumulation** Human waste remains and stacks up
Networking Nature builds self-organizing systems where everything is linked together	**Linear systems** Man-made devices so not follow a cycle, but have a 'start' and an 'end'
Imperfection Developing through attempts and errors. Imperfection is hard to manage but structures are more durable. Rigidity is sacrificed for better energy absorption and shock control. Geometries are complex and versatile	**Perfection** Technology strives for perfection. With little or no margin for adapting, it becomes fast obsolete Structures resist to high loads, but are generally not designed for absorbing energy Geometries are more simple to calculate

role in nature, and appears to be incompatible with Newtonian and quantum dynamics, which are reversible deterministic theories (Reichl et al. 2004). The deterministic description of reality recognizes all processes and events as caused by antecedent events and conditions in the respect of the physical laws of nature (Hoefer 2010).

6.1 Source and Synergy

An ecosystem can be described from an energetic point of view as a balanced system of constant energetic exchange between its parts. Superior and subordinated systems (ecosystems, organisms etc.) are synergetic and interdependent.

The energy for all life on earth comes from the sunlight shining on us, which is through plant's photosynthesis trapped and transformed into sugars, which are the base for all energy exchange inside and in between the living systems. The plants are at the first stage of energy-transfer, animals in their turn absorb the 'solar energy' through feeding first on plants (herbivores) and eventually on other animals (predators) (Nachtigall 2000).

As energy systems tend to evolve hierarchically to a greater complexity, the energy pattern is subject to alterations to keep the balance: in case of big growth, the system must either reorganize itself into a more complex structure with higher energy flow, which is more difficult to uphold, or collapse.

6.2 Metabolism

All organisms must not only produce energy but also convey it. It is, in fact, the energy flow through the system that maintains the organism alive. There is a strong relationship between energy, lifespan and body mass in living organisms strongly characterizing their body plans. If 1 g of living tissue consumes approximately the same amount of energy in its life independently of the organism, it is logical that bigger organisms expend more energy than small ones. From a metabolic point of view, this means that small organisms are metabolically more active, whereas large organisms have a slow metabolism and usually live longer. Unicellular organisms have an extremely efficient metabolism, as their volume is, theoretically, the most contained as possible in comparison with the large surface area.

Metabolisms are differentiated by regime of thermoregulation (Schmidt-Nielsen 2004).

- Constant body temperature (used by warm-blooded animals, as mammals) is a very demanding energy regime. It allows the organism to remain active in cold environments though few survive in very hot conditions. Very little of the food is converted into body mass, as most of it is turned into energy.
- Variable body temperature regimes (used by cold-blooded animals as reptiles) are less energy demanding. The organisms can survive during long periods without food and therefore the research is not so active. Few survive in cold climates and most of the food is converted into body mass.

During locomotion a biological system has two major strategies to improve its work-energy balance (Nigg et al. 2000).

- Storing and returning energy by deforming a structure (as a spring);
- Minimizing loss of energy. In other words, minimizing the use of energy that is not related to the performance but would be available to enhance the performance if it had not been spent unnecessarily (Goldspink 1981).

7 Motion Versus Change

7.1 Motion as Change on a Specific Time-Scale

Movement or change in Nature can be found in three different time-related forms (Zuk and Clark 1970):

- Slow change—As is Evolution, acts on a very long timescale. In a biomimetic perspective, this kind of change is rather analyzed to allow a deeper comprehension and optimization of the system's design.
- Moderate change—As for growth and decay, which embraces the lifespan of an organism. In the biomimetic perspective, these aspects are of interest when taking into account the grey energy or the life cycle of a given man-made product.

The two above-mentioned timescales of change are not appropriate models in the context of designing motion, for three main reasons:

1. The timescale required for obtaining any change is too vast to allow any perception and active reaction to the change. For a change to be rapid enough to classify as motion, it must be perceived.[3]
2. Change of form through Evolution, and hence also through growth and decay, implies creation (through multiplication or addition) and destruction of material, implying obvious technological limits.
3. Change is not reversible and hence not repeatable, while motion is based on the repeatability of the physical change.

In order to address motion, biomimics must therefore turn to a specific timeframe:

- Fast change—As through muscular, hydraulic, pneumatic action. This change is instantaneous, and achieved by internal or external manipulation. This category of change is in the center of the following chapters and sets the basis for a direct biomimetic study in view of a translation of some systems into a technological context.

[3]Human perception of speed is measured by the time taken for an object or event to cross the 180° arc of vision. Anything moving faster than approx. 45°/s starts to blur, while any object unable to trigger a stimulation of the eye after 2–3 s is perceived as stationary (Borst and Egelhaaf 1989).

7.2 Tropic and Nastic Movements

Plants are of extreme interest in the context of biomimetic kinetic design, as the mechanisms are generally less complex than other organisms in their geometry and mechanics, although many of the reactions are still carried out through complex chemical reactions.

Tropism indicates a slow adaptive motion of plants, generally achieved through growth, in response to different environmental triggers where the direction of the motion is strongly dependent on the direction of the stimulus.

- Phototropism reacts to *light*, as plants grow in the direction of the light source;
- Geotropism reacts to *gravity*, where roots and stem grow in opposite directions, when possible the first downwards and the second upwards;

More rapid tropic movements use activating and control mechanisms as pneumatic and hydraulic pressure to achieve higher flexibility:

- Heliotropism reacts to the *motion of the sun*, as in Sunflowers, which move by twisting the stem;
- Thermotropism reacts to *temperature*, as plants grow towards or away from heat sources;
- Hydrotropism reacts to the presence of *water*, as when plants' roots grow seeking water;
- Haptotropism or Thigmotropism react to *touch*, as when Passiflora tendrils search for support to grip and climb.

Nastic movements are faster and differ from tropic movements as they concern non-directional movement in response to a stimulus.

- Photonastic and Nyctynastic reacts to *light* (or darkness), as the leaves of plants fold when light levels are low;
- Hydronastic reacts to *humidity*, as for grass leaves which fold and curl up in wet weather conditions;
- Thermonastic reacts to *temperature*, an example are plants' leaves that can change shape depending on the cold or hot weather;
- Thigmonastic, haptonastic or seismonastic reacts to *touch*. The *Mimosa Pudica* is famous for closing its leaves when touched, as also the *Venus Flytrap* closes on its prey when adequately stimulated.

Out of these categories, mostly nastic movements are of interest in the present context and will be further explored.

7.3 Locomotion

Whereas most plants achieve motion while being anchored to a fix support, locomotion used by animals implies the capacity to change location through active or passive motion. The greater complexity and reactivity of animal structures adds many more parameters to their movements, together with a multitude of possible combinations, of which some are transversally recurring.

The *medium of locomotion* strongly defines the physical characters of the animal. All animals are perfectly adapted to the medium they move in, and many have evolved the capacity to do so in more than one medium: frogs for instance, have evolved limbs for terrestrial locomotion and webbed feet for aquatic locomotion. Most of these multifunctional organisms do not however have full motive efficiency in both media: where more functions need to coexist, full optimization does not appear to be possible. Also, many animals moving in two media have separate sets of tools (limbs or appendages in most cases) for each media: flying fishes possess both wings, feet and tail while birds have both wings and legs (Zuk and Clark 1970).

In the matter of terrestrial locomotion, most animals have evolved motion based on the displacement of the opposites of the body, either limbless (as serpentine locomotion or concertina) or with limbs (based on the motion of an even number of limbs).

In locomotion in fluids, water and air, most organisms rely on the action of an airfoil, or a broad surface as a fin or a wing, which is rhythmically pressed towards the media to transform muscular force into a vectorial movement. In fluids, a fixed surface simply oscillating back and forth does not provide any resultant movement as the resistance of the medium is equal in both directions. Therefore, the surface must adapt its geometry to different configurations to make sure to obtain thrust forward during the beat movement and no resistance during the recovering movement. The ability of molding the surface also requires the capacity to perform many different maneuvers as ascension or descension, acceleration and deceleration and not least different cruising and turning manoeuvers (Zuk and Clark 1970).

Movement and simple *control of movement* can be integrated in an organism's structure by considering the global interdependence of all structural elements. There are three main ways to control a kinetic transition process in a predictable and safe manner. In biological organisms, these possibilities are most often combined.

1. *Control of the joints*. Joints allow movement in definite directions while providing stiffness to other directions providing a base resistance for deformations. Joints may also provide some friction to resist unwanted deformations and thus allow movement control (Stein et al. 2000). Movement is introduced when applied forces act on a number of joints, which allow a change in geometry through different rotating movements.
2. *Geometrical arrangements*. The geometrical use of the curvature which is an encumbrance in bearing axial loads, can however be of interest to facilitate a more flexible and predictable arrangement. Long bones in vertebrates, as in particular leg bones are slightly curved with a geometry known as Euler buckling, which is allowed to bend when stressed, acting as shock absorbers, eventually breaking to dissipate even more energy (Gruber 2011). The force resisted is low but the strain energy absorbed is manifold higher than if the bones were straight: nature prioritizes security before performance.

3. *Boundary conditions*. Predictable structural limit conditions can be used to the system's advantage, as in the case of the Venus flytrap. This carnivorous plant uses the instantaneous geometric deformation of its leaves from convex to concave to achieve an extremely fast closing movement able to trap its preys (see the case study).

8 Conclusion

Nature, as technology, adapts by intervening on two different scales of design: optimization of the process on one hand and accumulation of single specific optimized solutions on the other.

Evolution achieves adaptation and optimization *through process* by:

- Cumulative instead of single step selection;
- Creation through mutation and destruction through selection.

While adaptation and optimization *through single solutions* is achieved through:

- Categorization and organization of the information;
- Developing of recurring patterns;
- Developing of Body Plans and primitives.

Supposing that each type of motion which an organism can create in a closed system of physical conditions (the Earth) would optimally work with a very specific combination of physical, mechanical and geometrical features, resulting in a maximum output with a minimum energetic input (in all forms), then adopting this particular combination of features would be of enormous evolutionary advantage.

For the purpose of facilitating a biomimetic solution transfer from biology to technology, the following chapters will proceed with a systematic classification of bio mechanic principles and solutions according to recurring typologies and patterns of motion, with the idea that all working systems can be organized as specific variations of combinations of given parameters.

The aim is to achieve a broad knowledge concerning both the most common parameters and the possible efficient combinations of these in a motion-related context.

References

Alexander C, Ishikawa S, Silverstein M (1977) A pattern language. Towns, buildings, construction, 1st edn. Oxford University Press, New York

Arthur W (1997) The origin of animal body plans: a study in evolutionary developmental biology. Cambridge University Press, Cambridge UK

Art Science Museum (2017) Human + the future of our species. Singapore August–October 2017

Borst A. Egelhaaf M (1989) Principles of visual motion detection. Trends in Neurosciences, vol 12
Dawkins R (1996) The blind watchmaker, why the evidence of evolution reveals a universe without design. W.W. Norton & Company, Inc., New York. ISBN 0-393-31570-3
Fuller RB, Applewhite EJ (1982) Synergetics, explorations in the geometry of thinking (n.d.). Macmillian, New York
Goldspink G (1981) The use of muscles during flying, swimming, and running from the point of view of energy saving. In: Day MH (ed) Vertebrate locomotion, symposia of the Zoological Society of London, nr.48, Academic Press Inc., London, UK. ISBN 0-12-613348-4
Gruber P (2011) Biomimetics in architecture, architecture of life and buildings. Springer-Verlag, Wien. ISBN 978-3-7091-0331-9
Hoefer C (2010) Causal determinism, stanford encyclopedia of philosophy. Accessed 2 March 2015 from http://plato.stanford.edu/entries/determinism-causal/
Latthe SS, Terashima C, Nakata K, Fujishima A (2014) Superhydrophobic surfaces developed by mimicking hierarchical surface morphology of lotus leaf. Molecules 19(4):4256–4283
Lynn G (2004) The future of architecture and philosophy. Public open video lecture, Saas Fee, Switzerland: Faculty of European Graduate School EGS, Media and Communication Studies, from http://www.youtube.com/watch?v=6KURvYOjwO4
Lynn, G. (2008). Form. Rizzoli International Publications, Inc
Madrazo Agudin L (1995) The concept of type in architecture, an inquiry into the nature of architectural form. Dissertation ETH Zurich No. 11115, from http://ecollection.library.ethz.ch/eserv/eth:39939/eth-39939-02.pdf
Mandelbrot B (1983) The fractal geometry of nature. Freeman, New York
Menges A, Hensel M (2008) Morpho-ecologies. Architectural Association, London
Menon C, Murphy M, Sitti M (2004) WaalBots: miniature wall climbing robots for space applications. In: Proceedings of the international astronautical congress, Vancouver, Canada, 4–8 October 2004. Accessed 27 September 2017 from http://nanolab.me.cmu.edu/publications/
Nachtigall W (2000) Das Große Buch der Bionik, Neue Technologien nach dem Vorbild der Natur, Deutsche Verlags-Anstalt, Stuttgart/München.
Nachtigall W (2003) Bau-Bionik, Natur – Analogien - Technik. Springer-Verlag, Berlin, Heidelberg, New York
Nigg BM, Stefanyshyn D, Denoth J (2000) Mechanical considerations of work and energy. In: Nigg BM, MacIntosh BR, Mester J (eds) Biomechanics and biology of movement. Human kinetics, Champaign, Ill
Phylogeny (2015) Merriam-Webster dictionary, Encyclopedia Britannica. Accessed 1 March 2015 http://www.merriam-webster.com/dictionary/phylogeny
Reichl L, Schechter R, Sudarshan G (2004) Report of the memorial resolution committee for Ilya Prigogine, University of Texas Austin. Accessed 2 March 2015 from http://www.utexas.edu/faculty/council/2003-2004/memorials/prigogine/prigogine.pdf
Schmidt-Nielsen K (2004) Scaling, why is animal size so important? Cambridge University Press, Cambridge, UK, (1 ed.1984). ISBN 0-521-31987-0
Stein R, Zehr EP, Bobet J (2000) Basic concepts of movement control. In: Nigg BM, MacIntosh BR, Mester J (eds) Biomechanics and biology of movement. Human kinetics. Champaign, Ill
Thompson D'AW (2006) On growth and form. Cambridge University Press, 1961 (Trans: Albus A). Eichborn AG, Frankfurt am Main
Valentine JW (2004) On the origin of phyla. The University of Chicago Press, Chicago and London. ISBN 0-226-84548-6
Wolpert L, Tickle C, et al (2011) Principles of development, 4th ed. Oxford University Press
Zinsmeister A (Hg/Ed) (2011) Gestalt der Bewegung/figure of motion, jovis Verlag GmbH, Berlin
Zuk W, Clark RH (1970). Kinetic architecture (n.d.). VanNostrand Reinhold, New York

Chapter 3
Geometry of Motion

Abstract Motion strongly depends upon the geometrical features of the elements it is made of, therefore in nature there is coevolution of morphology, movement type and movement control exploiting the morphological features also in energetics and mechanics. The chapter introduces six main geometrical parameters in locomotion, with a number of subcategories.

In nature, there is maximal exploitation of morphological features in energetic and mechanical contexts. Motion strongly relies upon the geometrical features of the body parts. This is achieved through co-evolution of morphology, movement type and movement control (Pfeifer 2000).

On geometry depend factors as inertia, weight and therefore energy absorption and balance. It's then not surprising that many body plan features reflect specific locomotory techniques (Valentine 2004).

The geometries analyzed belong to a variety of contexts, from the animal to the vegetal kingdom, and combine therefore aspects from a many parallel disciplines from taxonomy, biomechanics, biological scaling, etc. The aim is not to suggest alternative classifications to the state-of-the art in the respective fields, but to provide a framework for technological biomimicry purposes. The categories therefore present a fusion of aspects that are usually the subject of single specialized fields of study. The parameters identified are the outcome of a selection and simplification of characters that seemed relevant for the purpose. Many of the aspects can be found across scales and in many combinations (geometrical features for instance recur in combinations within the same organism), and should therefore be interpreted as a simplified record of variations.

S. Persiani, *Biomimetics of Motion*,
https://doi.org/10.1007/978-3-319-93079-4_3

1 Symmetry

Symmetry can be seen as the morphological translation, or the specular reproduction, of a unitary structure with an 'immunity to change with respect to a specific aspect': the human body is symmetric through its mid-plane with respect to appearance (Rosen 2008). As a physical expression of a defined set of behavioral rules, the symmetrical proprieties[1] of an organism are essential components in the management of information and growth of complex cellular systems. Total or partial symmetry in a body is the morphological foundation for its *reproducibility* and for the *predictability* of its structure.

Reproduction is achieved through duplication and differentiation of the body's smaller building blocks, the cells. Hence the repetition of approximate symmetries. *Predictability* of an organism's development and mechanics is the *conditio sine qua non* of any evolutionary process. Symmetrical proprieties are among the main factors defining an organism's body plan and are highly conservative features in evolutionary history (Valentine 2004) (Fig. 1).

Most organisms in nature, whether plants or animals, exhibit some kind of symmetrical or geometrically organized body plan which is the outcome of a polarization process that occurs in the earliest stages of the developing embryo. It defines the major body axes by symmetrically breaking up. The evolution of the existing variations in morphological geometry, possibly originating from a common cylindrical ancestor, are thought to be the outcome of the repeated evolutionary 'tinkering' of features under environmental constraints (Manuel 2009).

Correlation between morphological symmetry and locomotive efficiency has been indirectly demonstrated in biological and technological case studies: organisms with a higher degree of asymmetry tend to exhibit lower locomotive efficiency than symmetrical organisms. Even radially symmetrical organisms as jellyfish and sea cucumbers have evolved directional locomotion along their oral-aboral axis (Astley 2012). Asymmetrical features present several handicaps as causing more erratic and deviating movements, resulting in energetic efficiency as more force needs to be applied to the actuators to counterbalance the asymmetry, consequently achieving lower locomotive effectiveness and path efficiency (Wolpert et al. 2011; Gruber 2011; Valentine 2004; Bongard and Chandana 2000; Arthur 1997).

1.1 Bilateral Symmetries

Bilateral symmetry mirrors the overall body plan along the sagittal plane[2] dividing the body vertically into left- and right-halves with most organs and limb pairs on

[1]In this context, 'approximate symmetries', as a softer definition of the dichotomy between exact symmetries and asymmetries, are considered.

[2]General Anatomical Planes: *Sagittal*, dividing a body into left and right (oral-aboral or antero-posterior axis); *Coronal* (or frontal), dividing the body into front (ventral) and rear (dorsal);

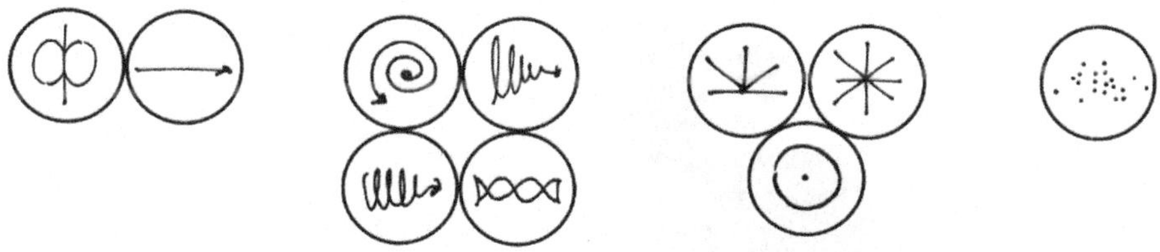

Fig. 1 Symmetrical features grouped in four categories (from the left and top to bottom). Bilateral (bilateral and linear); Revolving (spiral, conic helix, circular helix, double helix); Radial (bi-radial, spherical, radial); Asymmetric

either side. In most bilaterians the mirrored symmetry is particularly pronounced, with head and tail, back and belly clearly visible, while for others the polarity is restricted to the internal structures (as for *anthozoans*) (Manuel 2009). Bilaterianism in biological organisms is believed to have evolved once, becoming a permanent feature of higher animal species (Wolpert et al. 2011; Valentine 2014).

Example	1. Butterflies' and Moths' wings (*or. Lepidoptera*)

Linear symmetry can be considered a sub-category of the bilateral symmetry. It suggests the organism is mirrored along the sagittal plane, and partially along the transverse plane, reproducing a sequence of similar elements along the central body axe (Valentine 2004) (Fig. 2).

Example	1. Snakes (sub. *Serpentes*) 2. Earthworm (sub. *Lumbricina*) 3. Pea pods (fam. *Fabaceae*)

Bilateral locomotion is usually thought of as being linked to directional locomotion along the anterio-posterior axis of the main body length, with the dorso-ventral axis oriented with respect to gravity. This may in part explain the omnipresence of bilaterians in the terrestrial environment, where gravity has a much stronger impact than in water. Many other bilaterians (as mussels, barnacles and tapeworms) however have lost their directional locomotion while maintaining the mirrored body plan. It would therefore seem that the adoption of specific symmetries is not constrained by the biological functions of the organisms, but are more the outcome of a morphological evolution under multiple external pressures (Manuel 2009).

Transverse (or horizontal), dividing the body into a superior and inferior part (see also chapter on representation of motion and orientation).

Fig. 2 Bilateral symmetry in a tortoiseshell cat, highlighted by the two-sided color pattern of its coat; linear symmetry in a snake

1.2 Revolving Symmetries

Revolving symmetries are a way to preserve the proportions of an organism during growth, an advantage in terms of energy and information minimization, and are therefore often found in animate nature (Wolpert et al. 2011; Valentine 2004) (Fig. 3).

Spiral symmetries develop bi-dimensionally, along a single plane and around a fixed central point/axis individuated outside the physical boundaries of the organism, at a continuously increasing or decreasing distance from the point. Among the most common spirals in nature, the Archimedean, Fibonacci and logarithmic spirals.

Example	1. Butterflies' proboscis (or. *Lepidoptera*)—spiral of Archimedes 2. Central disc in Sunflowers (gen. *Helianthus*) and in Daisies (gen. *Argyranthemum*)—Fibonacci sequence spirals 3. Nautilus' shell (fam. *Nautilidae*)—Logarithmic spiral

Helixes develop three-dimensionally, revolving around an axis at a variating or constant distance.

A *circular helix* has a constant radius and constant torsion.

Example	1. Feather duster worm (gen. *Bispira*) 2. Passion flower's tendrils (gen. *Passiflora*)

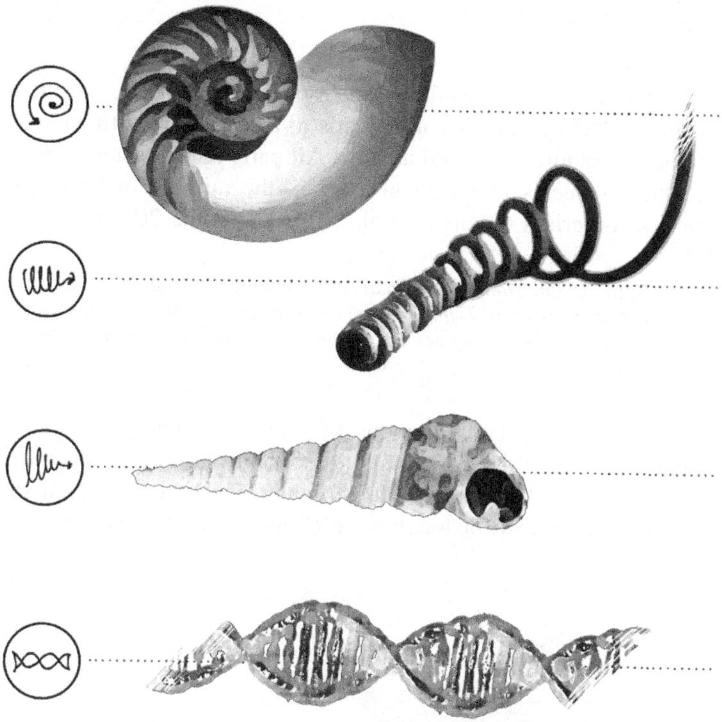

Fig. 3 Types of revolving symmetries. From the top: spiral symmetry in the cross-section of a Nautilus' shell (gen. *Nautilus*); circular helix symmetry in Passion flower's tendrils (gen. *Passiflora*); conic helix geometry in a Turritella sea snail shell (gen. *Turritella*); double helix geometry found in DNA structure

The *conic helix* will develop with a progressive variating distance from the axis.

Example	1. Passion flower's tendrils (gen. *Passiflora*)

The *double helix* consists of two congruent and parallel spiraling helices on the same axis.

Example	1. DNA molecule

1.3 Radial Symmetries

In *radial symmetries*, the organism develops a multimeric, multilateral and equidistant shape. The symmetry corresponds to a repetition (*n* times) of the same set of anatomical structure(s) around an axis. All parts face the same environmental conditions. The *n* value is indicated in a prefix the number of limbs: tri-, tetra-, penta-, hexa-, octa-merism (Wolpert et al. 2011; Valentine 2004).

Example	1. Apple (spec. *Malus domestica*) has internal pentamerism 2. Clover leafs (gen. *Trifolium*) are trimeric and eventually tetrameric 3. Corals and sea anemones (cl. *Anthozoa*) are hexa-/octameric through their tentacles

The physical structures and functions of radial symmetries are of extremely diverse natures: from the hydrodynamic functioning of the syconoid sponges' aquiferous organization, the light skeletal structure of the hexactinellid sponges, to the use of ciliated structures for locomotion (Manuel 2009) (Fig. 4).

Fig. 4 Types of radial symmetries. From the top: radial symmetry in a pentameric Sea star (cl. *Asteroidea*) the number of arms depend on the species and can go from five up to fifty (as in the *Labidiaster annulatus*); bi-radial symmetry in a Bee orchid (spec. Ophrys apifera); spherical symmetry in Volvox green algae (gen. Volvox)

Cylindrical symmetries are considered as a variation of a radial symmetry developed around a single axis with and infinite number of planes of reflection passing by that axis.

Bi-radial symmetry can be considered as a specific variation of a radial symmetry with $n = 2$. Bi-radial systems are differentiated from bi-lateral systems at a taxonomic level as the body has only one polarity axis (and not two) and develops true multimerism with a set of identic and superimposable structures around the axis (Manuel 2009). The body components are arranged radially with a succession of different elements, intermittently symmetrical.

Example	1. Comb-jellies (phyl. *Ctenophora*) 2. Bee orchid (spec. *Ophrys apifera*)

Spherical symmetry, the radial development occurs in three dimensions. The organism can be cut into two identical halves through any cut that runs through the geometrical center. Spherical symmetry is quite common in unicellular and in a few multicellular systems, while it is almost inexistent among developed metazoans (Manuel 2009).

Example	1. Volvox green algae (gen. *Volvox*) 2. Demosponges (gen. *Tethya*) display partial spherical symmetry

1.4 Asymmetry

The concept of 'asymmetry' can assume very different meanings depending on the context. Almost no organism in nature shows perfect symmetrical proprieties as even specular organs and segments develop independently and may specialize on different tasks. The proprieties further considered are the intrinsic physical characters specific for each species, not the exceptions or small differences due to the development conditions, but as elements directly characterizing the species' morphology or with a direct functional asymmetrical propriety (Wolpert et al. 2011; Valentine 2004) (Fig. 5).

Partial asymmetry, dissymmetry, or secondarily disrupted symmetry, as the outcome of a polarization process, is extremely common and can be found in most organisms. In partial asymmetry, sometimes even used as a synonym of bi-radiality, the organism presents asymmetric proprieties in one dimension, usually directly connected to a functional propriety of the body part.

Example	1. Fiddler crabs (gen. *Uca*) have one big and one small chela (pincer) 2. Most primates (or. *Primates*) have opposable thumbs

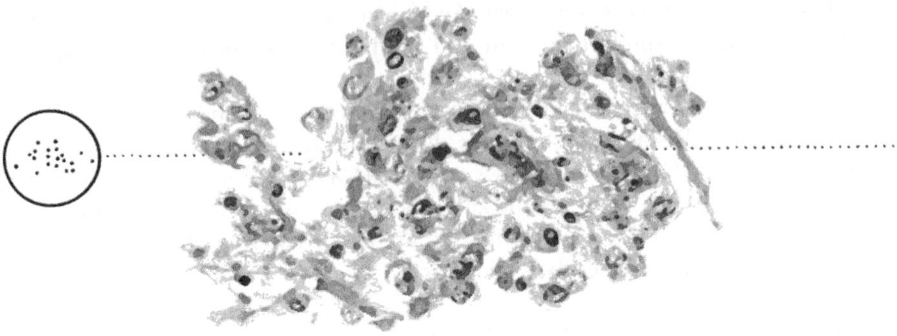

Fig. 5 Total asymmetry in a syncytium

Total asymmetry, in the sense of morphological indefinition, is quite uncommon, and can be found in organisms that present asymmetric proprieties in three dimensions with no possibility to find any kind of symmetry in the overall body.

Example	1. Sponges (phyl. *Porifera*) 2. Syncytia structures (ex. Sp. *Trichoplax adhaerens*)

2 Structure

Both unicellular and multicellular animals achieve motion by combining rigid compression-resistant and flexible tension-resistant elements. The skeletons' design radically influences force transmission, function and overall organization. It is thought that skeletons evolved either for protection or support, although evolutions has probably exploited both as long as they did not interfere with one another (Kier 2012; Biewener 2007; Valentine 2004; Wilson et al. 2000; Elder and Trueman 1980) (Fig. 6).

Fig. 6 Types of body structures (from the left): endoskeleton, exoskeleton and hydrostatic skeleton

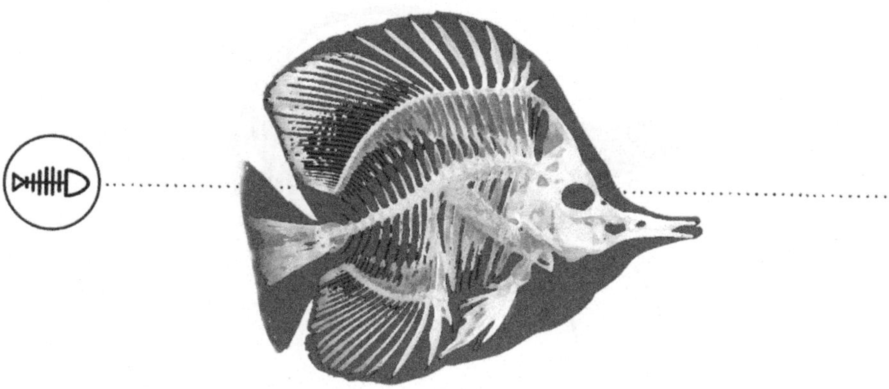

Fig. 7 Internal structural elements in a fish

2.1 Endoskeletons

Endoskeletons, 'internal skeletons', use internal stiff bony elements to support themselves and transmit forces, activated by muscles, which envelop the rigid parts and use them as levers. The rigid supporting structure is dimensioned to provide maximal support while contributing with little weight and have therefore a typically honey-comb like structure (Thompson 2006). The structure is designed to mechanically participate in the motion of the animal, and not as shielding element. As fractures are a significant risk for the organisms, vertebrate bones operate within broad safety margins. Bones are flexible enough (<1%) to allow substantial energy absorption before failure (their geometry and cartilage in the joints are energy absorbing and load distributing) and are able to heal (Biewener 2007; Valentine 2004) (Fig. 7).

Example	1. Vertebrates (Sub.ph. *Vertebrata)*

2.2 Exoskeletons

Exoskeletons ('external skeletons') are shell-like external rigid elements enclosing soft-bodied parts. These structures are excellent protective elements against predators, damage and desiccation but do however limit the growth of the animal: many must shed regularly to grow. Moreover external shells and protections are usually unable to heal once exterior damage has occurred (Biewener 2007; Valentine 2004) (Fig. 8).

Experimental compression tests on turtle's carapaces, which are thought to have evolved derivating from the endoskeletal ribs, showed a similar phenomenological stress–strain behavior as of that of metallic foams (Hirasawa et al. 2013; Damiens et al. 2012).

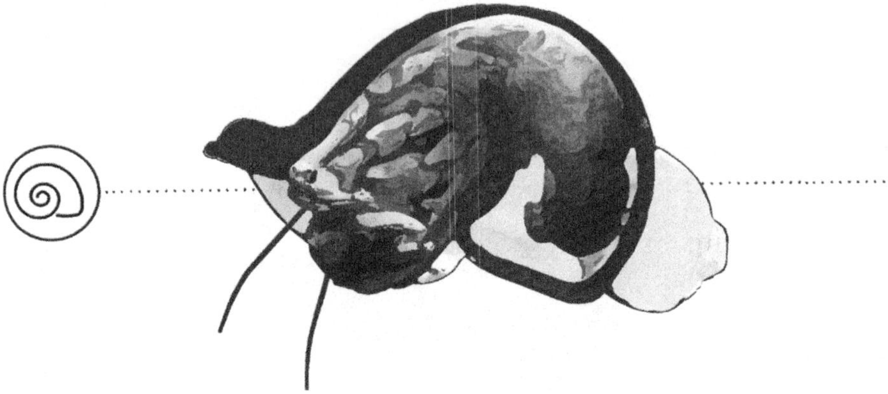

Fig. 8 Cross section of a Hermit crab (sup. fam. *Paguroidea*) retracting in a scavenged seashell. To protect its soft abdomen from predators, these crustaceans inhabit empty mollusc shells, vacating their old shell when a larger or better suited shell is found

Example	1. Grasshoppers (sub. *Caelifera*) 2. Clams (cl. *Bivalvia*) 3. Turtles (or. *Testudines*)

2.3 Hydrostatic Skeletons

Hydrostatic skeletons consist of a pressurized fluid-filled cavity surrounded by a tension-resistant fiber-reinforced wall activated by muscles. While endo- and exoskeletons enable organisms to transmit force with very little deformation, hydrostatic structures use the incompressibility of the internal fluid, and volumetric constance to obtain mechanical amplification by changing the organism's shape. The contraction of the muscles and the resulting decrease in one of the dimensions results in an increase in another dimension. By actively (with muscles) or passively (with connective tissues) controlling the various dimensions, a wide array of deformations, movements and changes in stiffness can be created. In most cases, the locomotion is slow and non-dramatic (Biewener 2007; Valentine 2004) (Fig. 9).

Most structures relying on hydrostatic skeletons have a cylindrical shape where the diameter is controlled by different muscles, arranged perpendicular to the longitudinal axis (circular musculature, radial musculature or transverse musculature), and length is controlled by longitudinal fibers. Deformations can occur at any point along the structure, unlike organisms with rigid skeletons (Kier 2012) (Fig. 10).

Crossed-helical fibers allow smooth bending movements, uniform and precise elongation and resist twisting and torsion (Fig. 11).

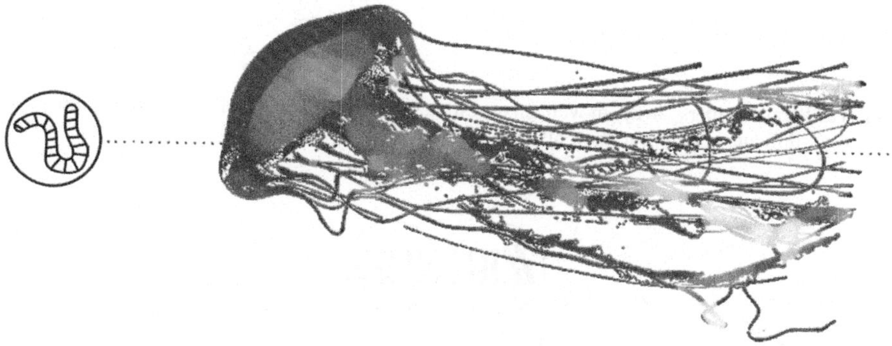

Fig. 9 A jellyfish (sub. Ph. Medusozoa) is a typical example of a soft-bodied animal

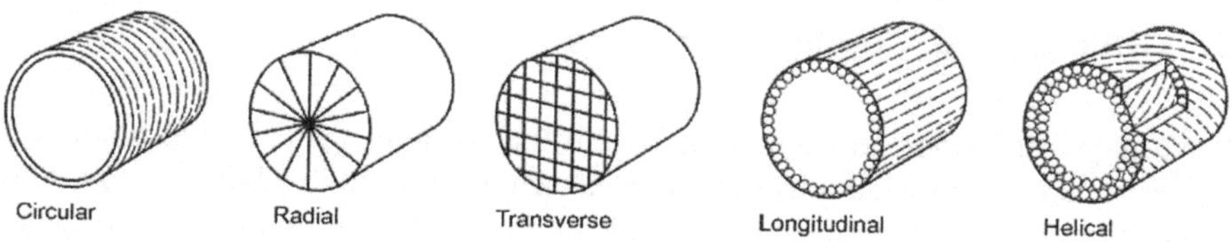

Fig. 10 Cross section of muscle fiber orientations in hydrostatic skeletons. Circumferentially orientated fibers, radial and transverse are used in elongation and resistance of bending movements. Longitudinal fibers shorten and bend the body. Helical fibers, arranged both in clockwise and counterclockwise direction, achieve torsion and twisting in muscular hydrostats (from Kier 2012)

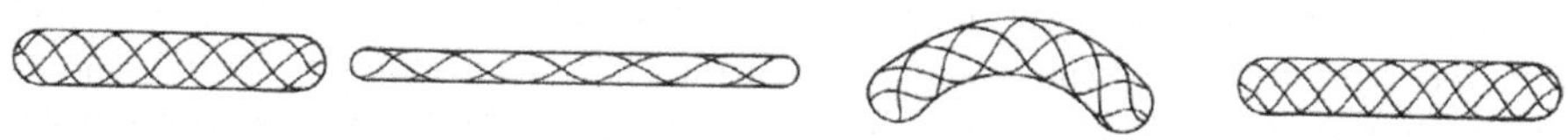

Fig. 11 Crossed-helical fibers array movements: resting position, extension, smooth curving, torsion resisted (from Kier 2012)

Example	1. Polyps 2. Vermiform animals, earthworm (sub. *Lumbricina*)

Parallel fiber arrays are arranged orthogonally and are able to resist length change and bending movements, allowing torsion (Fig. 12).

Muscular hydrostats lack the fluid-filled cavity, and consist instead of densely packed arrays of muscles and fibers, arranged so that all three dimensions of the structure can be actively controlled allowing precise complex bending movements. Contraction in one direction causes elongation of the opposite muscle bundles.

Example	1. Tongues 2. Jellyfish (s.phy. *Medusozoa*)

Fig. 12 Parallel fibers array movements: resting position, no extension allowed (orthogonal stiffness), difficult bending, torsion allowed (from Kier 2012)

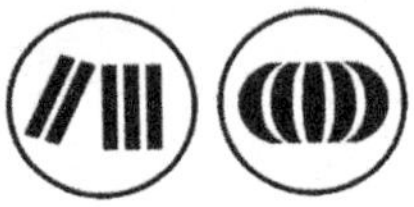

Fig. 13 Types of change considered (from the left). Change in orientation and change in geometry

3 Dimensional Change

For a physical change to occur in a system that is defined by a fixed amount of matter, thus excluding any physical changes achieved through addition (growth) or subtraction (decay), there must be a change in the orientation or in the organization of the system or of its parts. In the field of mechanics, dynamics can be subdivided into mechanics of rigid or deformable bodies.

As seen in the previous point, three types of 'structure' can be found in nature: endo- and exoskeletons are based on the mechanics of rigid systems and hydrostatic structures are based on deformations. The type of dimensional change, *orientation* and *geometry* respectively, are therefore deeply rooted in these types of structures. Most organisms however, set apart the mechanics of their overall body structure, integrate a multitude of sub-systems based on different principles (Homo sapiens has an endoskeleton, bound to the mechanics of rigid systems and change through orientation, while boneless muscles as the tongue can deform and change geometry) (Fig. 13).

3.1 Orientation

Change in *orientation* is typically used for bodies made out of rigid parts. A change in position occurs when the organism or its parts changes from a conformation A to a conformation B due to a rotation, a translation or both, independently of any dimensional change (Moloney 2011).

Planar motion. The kinetic elements are structurally and dynamically acting in a single plane, and motion can be described in a bi-dimensional coordinate system (Fig. 14).

Example	1. Clam (cl. *Bivalvia*), closing and opening its shell

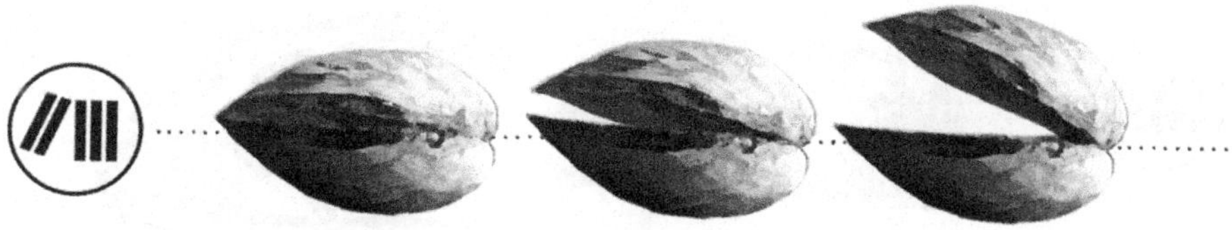

Fig. 14 Planar change in orientation in a Clam (cl. *Bivalvia*)

Spatial motion implies that the motion's geometry must be described in a three-dimensional coordinate system or with non-Euclidean geometry.

Example	1. Many kind of antennae

3.2 Geometry

A change in *geometry* can occur only in bodies or parts of the body that are made of flexible or soft-bodied parts. The movement is connected to a dimensional change due to a *deformation* of the body in at least one of its entities (Moloney 2011).

Linear change is achieved in one dimension. The body hence has variable length, while width and height are fix, or the variation is marginal.

Example	1. Snail (cl. *Gastropoda*), stretching out its antennas

Surface change is bi-dimensional. The body has variable length, variable width, and fix or marginal change in height.

Example	1. Fish skin is highly elastic, it allows the fish transmit the locomotion's' oscillatory movements along the body, acting as a spring

Volume change is three-dimensional. The body can change its length, width and height (Fig. 15).

Example	1. Fregatebird (gen. *Fregata*) blowing up its pouch

Fig. 15 A recently discovered species of frog, the Mutable rain frog (sp. *Pristimantis mutabilis*), has the capacity to change its skin texture from rough and spiny to smooth in a few seconds (Science & Children 2015)

Fig. 16 Types of structural subdivision (from the left). Monobloc, segmentation and appendages

4 Cleavage

Cleavage, in biology, is a series of rapid cell divisions without growth that divides the embryo up to a number of cells. In the present context though, since most examples taken into consideration are in the range of adult organisms—and thus are complex systems of trillions of cells—cleavage is intended in the more technical sense of a physical macro-subdivision of a body into multifunctional parts (Wolpert et al. 2011; Valentine 2004; Biewener 2007; Arthur 1997) (Fig. 16).

Cleavage is an essential feature in most body systems, enabling not only multifunctionality but also locomotion. In this context, the parts involved in kinetic actuation can be described as adopting one or more of the following cleavage features: monobloc, segmentation, appendages.

4.1 Monobloc

A *monobloc* body-system can be considered to be formed by a single integral body-unit. Monobloc organisms or systems achieving locomotion are, by logic, flexible or soft-bodied as any rigid parts need a minimum of one degree of freedom to move (Fig. 17).

Example	1. Flatworm (Phy. *Platyhelminthes*)

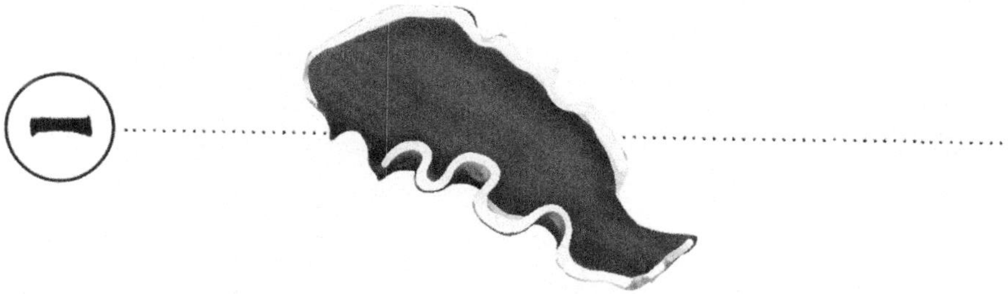

Fig. 17 Flatworms (Phy. *Platyhelminthes*) are simple, unsegmented bilateral invertebrates. They have no body cavities, no circulatory or respiratory system and must rely on nutrients and oxygen to diffuse through their flattened shapes (Barnes 1998)

4.2 Segmentation

Segmentation and 'metamerism' describe, in biology, a division of an organism's body-parts into a series of morphologically similar units or segments. In this context, the concept of segmentation will be applied to any morphological subdivision of a single body-part into a repetition of sibling-geometries, contributing to locomotion. Depending on the grouping and functions carried out by the body-segments the segmentation can be considered to be either homonymous (strict serial succession of the parts), or heteronomous (segments that perform similar tasks are grouped together) (Fig. 18).

Example	1. Vertebral spine—Homonymous segmentation 2. Earthworm (sub. *Lumbricina*) internal subdivision—Homonymous segmentation 3. In arthropods (ph. *Arthropoda*), the segments are grouped into head, thorax and abdomen parts ('tagmata')—Heteronomous segmentation

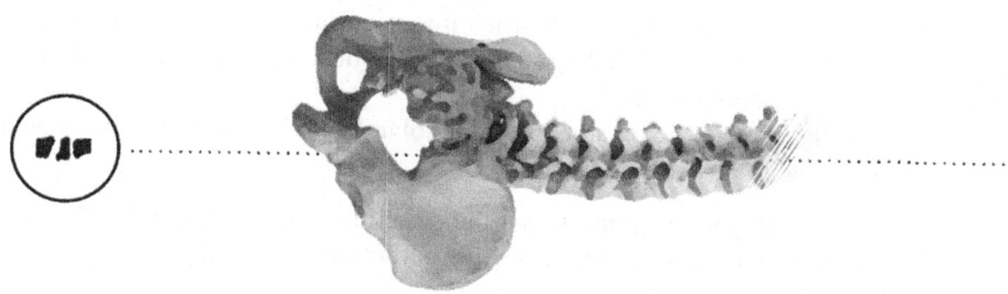

Fig. 18 Lower part of the human vertebral spine. The hononymously segmented bone elements allow flexibility as well as support to the body

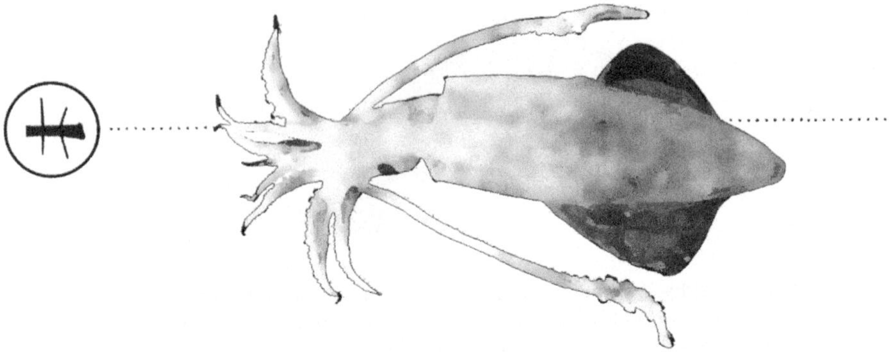

Fig. 19 Appendages in a Squid (Phy. *Platyhelminthes*): two tentacles, eight arms and two fins

4.3 Appendages

Appendages are, in biology, homologous body parts, or natural prolongations, that extend from a body segment. Appendages can be of rigid or of flexible nature, and generally have the function to maximize the interactive, sensorial and physical capability of an organism, and have in most cases dedicated motor systems (Fig. 19).

Example	1. Limbs and tail 2. Antennae and mandibles

5 Scale

Changes in size in otherwise similar animal and vegetal structures affect the structure as well as the functioning of the systems as a whole. In sibling structures of different size, linear, areal and volumetric dimensions are scaled at different rates. This affects all processes depending on those entities: musculoskeletal design, energy balance, locomotion, etc.

Dimensionally, three basic parameters are involved when adapting structures to a change in size (Schmidt-Nielsen 2004):

- Change in dimensions or proportions: the structure is made thicker or thinner, shorter or longer than a reference sibling structure (ex. Comparing the limb bones of a cat and a cattle);

Fig. 20 Scales considered (from the left): micro, small, medium, large, macro

- Change in materials: which are less or more rigid, lighter or more resistant (ex. Comparing ground beetles' thin and light wing-covers and a turtle's thick and hard carapace, that are analogous by the shape and protective function but differ in scale);
- Change in design, from compression to tension elements or changing the overall principles of locomotion (ex. while a unicellular organism might swim many times its body length using cilia, a shark covered in cilia would get nowhere).

As in engineering, for every feature and system, there is a defined size-range within which a particular solution maintains its efficiency. Changes in body size have extremely important consequences in locomotion. For organisms moving in fluids, size has great impact on the Reynolds number[3] allowing those with higher inertia to glide (as sharks and bigger birds), while smaller ones are held back by the fluid viscosity and move with bursts of activity (as shrimps and sparrows) (Biewener 2007; Valentine 2004; Wilson et al. 2000). Classifying biological organisms in terms of scaling involves not only physical dimensions but also factors as metabolic rates and locomotive parameters.

Also during growth, the features of an organism do not develop simultaneously or at the same rate. As it is impossible for an organism to produce another organism of its same size, a smaller offspring which in time grows up to the same size of the parent by changing scale must be produced. An isometric scaling (a proportionate square-cube change in dimensions) would eventually result in a mismatch between the organism's body scale and physical demands. Hence, allometric scaling (any scaling deviating from isometry and resulting in a change in proportions during growth), is common in most animal and vegetal systems (Damuth 2001) (Fig. 20).

In this context, the dimensional factors are divided into arbitral dimensional categories based on the organism's body mass in Kg [Mb], as the energy efficiency of the locomotion in biomechanics, the Cost Of Transport [COT], is expressed as a function of the body mass.

$$COT_{total} = 10.7\,Mb - 0.316$$

[3]Reynolds number (Re) is a dimensionless variable, a ratio of inertial forces divided by viscous forces, indicating viscosity and if the fluid currents are turbulent or laminar (Biewener 2007; Valentine 2004).

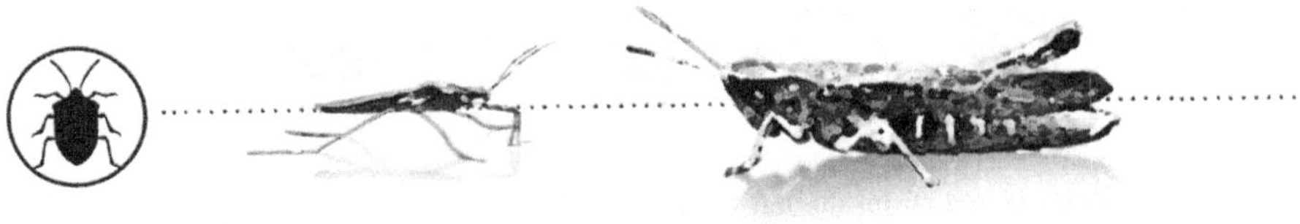

Fig. 21 Micro. To the left, a water strider and to the right, a grasshopper. These two examples show how the body features can change very drastically at a small scale: while the water strider has proportionally long legs and smaller body-mass to allow the water surface to support its weight, the grasshopper is more bulky and has developed strong hind legs to jump

5.1 Micro (Mb < 1 g)

At low Reynolds numbers viscous forces dominate, turbulence is absent. As symmetrical movements would cancel out each other, organisms rely on asymmetrical kinetics to achieve propulsion. Body shape is largely irrelevant to the ability to move through water or air, and is more bound to reduce friction drag by diminishing the body's surface area (see also flagellar and ciliary swimming in the sheet on limbless locomotion) (Biewener 2007) (Fig. 21).

Example	1. Water striders (fam. *Gerridae*)

Water striders take advantage of their small size to use surface tension of water to support themselves. Size range is very limited: if the animal is too large the water cannot support its weight, if it is too small it would not overcome the fluid's viscosity with its own inertia, remaining stuck to the surface (Biewener 2007).

5.2 Small (1 g < Mb < 1 kg)

The design of small animals show a big variation in shapes as the weight is not too much of an issue. Small sizes are diffused throughout many animal phyla (Schmidt-Nielsen 2004). Small terrestrial animals are characterized by stable postures (often crouched) and low centers of mass. Since stride length is a clue factor in speed, small animals have contained speeds compared to bigger animals. The curled up posture facilitates however maneuverability which more than compensates for the slow speed, allowing them to make sharper turns to escape their pursuer. The same can be said for birds: long and narrow wings allow gliding and soaring, small and stubby wings can most often not glide but are highly maneuverable (Biewener 2007) (Fig. 22).

Example	1. Bee hummingbird (sp. *Mellisuga helenae*) ~ 2 g 2. Rabbit (fam. *Leporidae*) ~ 105 g

Fig. 22 Small. Sibling geometries in a rabbit and a squirrel, same features (tetrapody, upright ears, fur, tail, strong hind legs, claws, etc.) but with different size to fully adapt the organism to a specific environment (running on ground vs climbing trees) and developing different locomotory features

5.3 *Medium (1 kg < Mb < 30 kg)*

Size-related shifts in posture, by running in more upright postures and aligning their joints more closely with the ground reaction force, are an important advantage in medium-sized animals to maintain specific structural safety factors during high speeds, reducing the force their muscles and bones must support, losing however part of the maneuverability (Biewener 2007) (Fig. 23).

Example	1. Cat (sp. *Felis catus*) ~ 4 kg 2. Kori and great bustrards (~13–16 kg) are the largest living flying animals. There is a size limit to muscle-powered animal flight: it is unlikely that a flying vertebrate over 25 kg has ever existed (Biewener 2007)

Fig. 23 Medium. Compared to the rabbit and squirrel in the previous category, this fox still presents similar features (tetrapody, upright ears, fur, tail, claws, etc.), but this time the posture is more upright and legs are longer

Fig. 24 Large. Compared to the fox in the previous category, this bear shows how large animals are comparatively more massive

5.4 *Large (30 kg < Mb < 300 kg)*

The larger the animals the more limited is their design: in animals built with the same 'building blocks' with similar proprieties (ex. Mammals), a large animal must be more bulky and have proportionally more massive limbs to support its body weight than a small animal (Fig. 24).

Example	1. Human (sp. *Homo sapiens*) ~ 70 kg 2. Brown bear (sp. *Ursus arctos*) ~ 130–200 kg

5.5 *Macro (300 kg < Mb)*

Very large animals have limited locomotive performance with respect to speed (Fig. 25).

Fig. 25 Macro. Setting the size of the adult animals apart, it is possible to see how the body features change according to the species. In this case, the body structure of a baby elephant is much more bulky than that of another animal of the same size, as for instance a big, full-grown dog

Example	1. Cattle (sp. *Bos primigenius*) ~ 420 kg 2. Elephant (fam. *Elephantidae*) ~ 7 T 3. Blue whale (sp. *Balenoptera musculus*) ~ 170 T

6 Motion Geometries

From a geometric point of view, motion can be seen as an oscillation between two or more morphologies: one complete cycle is achieved when the body returns to the initial position, allowing the set succession of changes to be repeated all over. A movement must therefore be reversible (growth is therefore not considered in this context) and must potentially work in a closed-loop system that can be divided into *n* temporal phases, from [t_0] (temporal starting point) to [t_n] (intermediate shapes assumed in *n* temporal units), returning to the initial position in t_0.

Specific motion geometries recur in reversible systems in nature (Kier 2012). By enabling selected movements through flexibility in one of their entities, these shapes can conserve kinetic energy in the structure and release it in the counter-movement (see also antagonist movement). These geometries are used broadly, and with very different purposes and outcomes, from locomotion to physical features. An undulating snake will change from a [t_0] S-shape to a mirrored [t_n] S-shape; a loop-walking Geometer caterpillar (fam *Geometridae*) from a linear [t_0] to an upside-down U shaped [t_n]; while a bird's feather flexes along its length to accommodate the linear flow of the air on the wing surface to reduce unwanted turbulence. The two first systems address aspects of locomotion but are radically different in size, structure and use different locomotion patterns, while the third one is one detailed feature of a greater organism (the bird). What these three movements have in common is their motion geometry, or the type of geometrical change between the temporal phases [$t_0 - t_n$] and [$t_n - t_0$], that are achieved through a succession of bending movements along the body (Fig. 26).

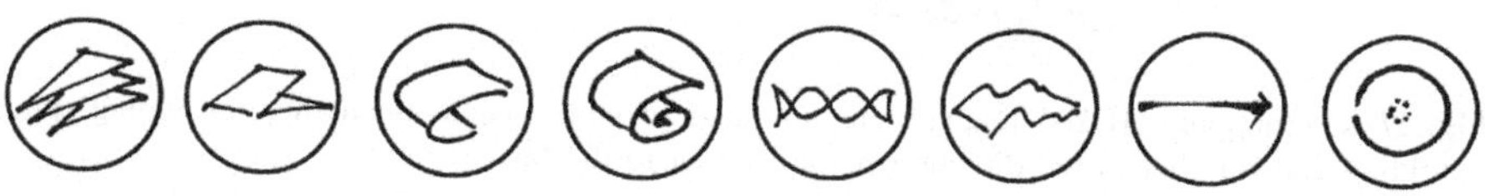

Fig. 26 Geometries of motion (from the left): overlapping, folding, bending, rolling, twisting, buckling, extension, expansion

Eight main motion-connected geometries have been identified in a vast number of systems (from vegetal to animal), and across many scales (from more to less complex organisms, and to parts of organisms and body features). All organisms integrate combinations of most of these geometries in a complex synergy of parts with kinetic potential. Motion is enabled or constrained depending on the symmetrical proprieties of the body. Bending, rolling, twisting and extension motions are typical of linear bodies. Flat and bi-dimensional systems prefer bending, rolling, twisting or buckling motions. Folding and overlapping is mostly used when the parts need to provide stability. Expansion requires spherical or balloon-like shapes.

6.1 Overlapping

Overlapping geometries are organized through a pre-determined superposition of parallel planes, which are in most cases identical copies of one basic shape (Fig. 27).

Example	1. Scorpionfly (or. *Mecoptera*) can reduce the length of its bosom by sliding the segments it is divided into under one another, like a monocular

Fig. 27 Overlapping. Birds' feathers are arranged to slide one over another, providing quick unfolding and stability in the direction perpendicular to the fan

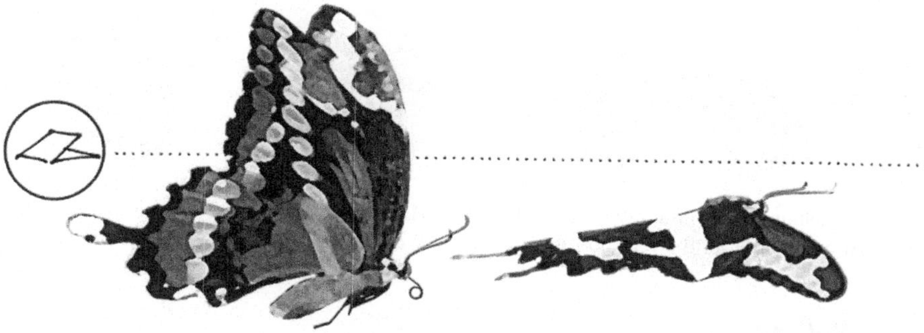

Fig. 28 Folding. Butterflies fold their specular wings following the oral-aboral body axis

6.2 Folding

Folding geometries organize surfaces using pre-determined folding schemes by dividing the surfaces in two or more segments and bending the planes so that one part covers another. Typically the planes are all linked to one another through common folding edges. The folding axes can be individuated inside the physical boundaries of the organism, creating more or less sharp folding edges (Fig. 28).

Example	1. Butterflies (or. *Lepidoptera*) opening and closing their wings

6.3 Bending

Bending geometries are closely related to folding geometries, with the exception that the folding axis can be individuated outside the physical boundaries of the organism, resulting in visibly curved edges instead of sharp ones (Fig. 29).

Example	1. Geometridae caterpillars (fam. *Geometridae*) have leg pairs at each end, and move by bending their length, grasping with one end while drawing up the other end

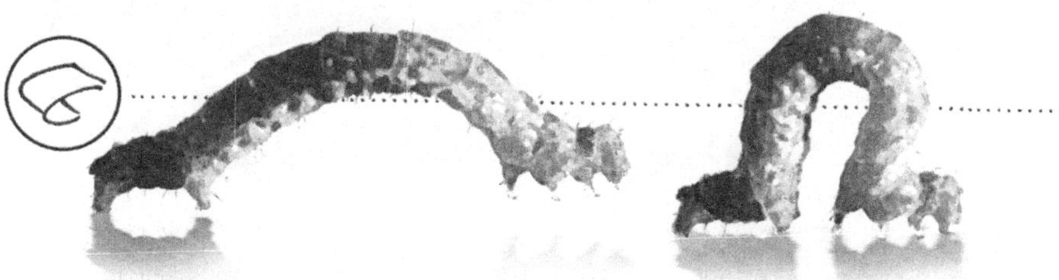

Fig. 29 Bending. A Geometridae caterpillar moving forward by bending along its body length

Fig. 30 Rolling. The commonly called pill-bug (sp. *Armadillidium vulgare*) rolls into a ball when disturbed, protecting all the legs and its underside by exposing only the hard exoskeleton

6.4 Rolling

Rolling geometries organize mainly mono-directional planes and volumes by rolling up the shape so as to obtain a compact geometry. The area in contact with the exterior is as small as possible, protecting the internal rolled-up elements. The geometry is particularly high-efficient from an organizational and technical point of view allowing the element to deploy in very short time lapses (Fig. 30).

Example	1. Butterflies' and Moths' proboscis (or. *Lepidoptera*) up-/un-rolls 2. Diplopoda (cl. *Diplopoda*) roll up spirally to protect their multitude of feet, organizing them along the inner perimeter of the spiral

6.5 Twisting

Twisting geometries are typically obtained in elongated structures where a flexible plane passing through the structure's rigid axis is longer than the axis itself, resulting in a deformation of the plane (Fig. 31).

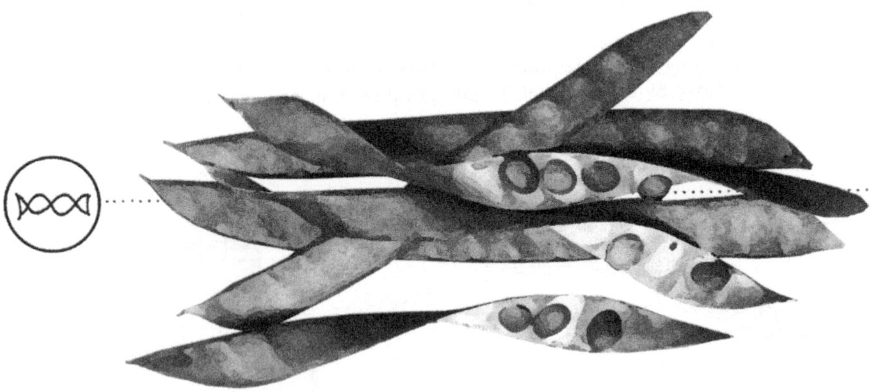

Fig. 31 Twisting. Many kinds of grasses and pods roll, curl and twist when the weather conditions change. In this case, the Bauhinia pea pods have twisted after desiccation, breaking up the pods and releasing the seeds

Example	1. Bird feathers flexing under air loads 2. Grass leaves can twist around in particular humid conditions

6.6 Buckling

Buckling, *undulating* or *crumpled* geometries organize mono and bi-dimensional surfaces by shortening the prevalent dimensional extension of the element by occupying as much bi-dimensional space as possible (Fig. 32).

Example	1. Mammalian skin, buckles in specific folding areas of the body

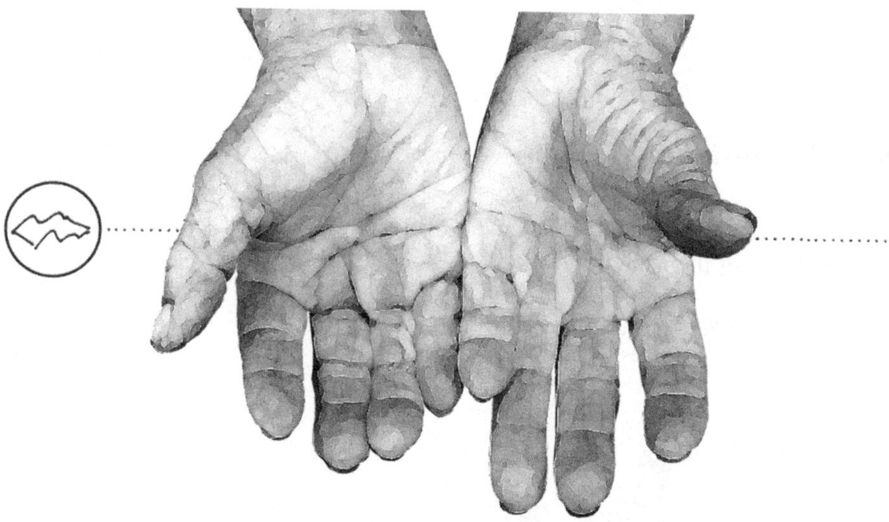

Fig. 32 Buckling and crumpled. Pre-determined folding- and wrinkle-patterns allow the skin on the underside of the hand fold following the changing size of the hand's surface

6.7 Extension

Extension and *compression* geometries are constant in volume, the movement depends on a modification in shape by adjusting their linear dimensions (length, width, height): contraction and decrease in one of the dimensions results in an increase

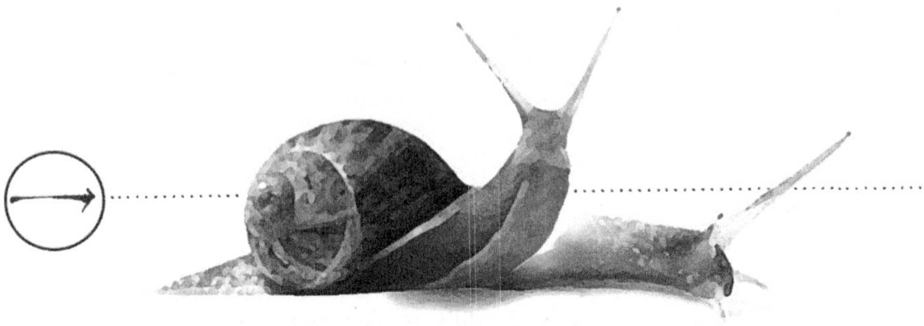

Fig. 33 Extension. A snail can extend its body and antennas with no apparent change in thickness of the body-part

in another dimension. By actively or passively controlling the dimensions, a wide array of deformations, movements and changes in stiffness can be created (Fig. 33).

Example	1. Hydrostatic skeletons

6.8 *Expansion*

Expansion geometries are characteristic of flexible structures, capable of dividing and multiplying their interior volume by many times (Fig. 34).

Example	1. Tick (sup.fam. *Ixodoidea*) 2. Pufferfishes (Fam. Tetraodontidea)

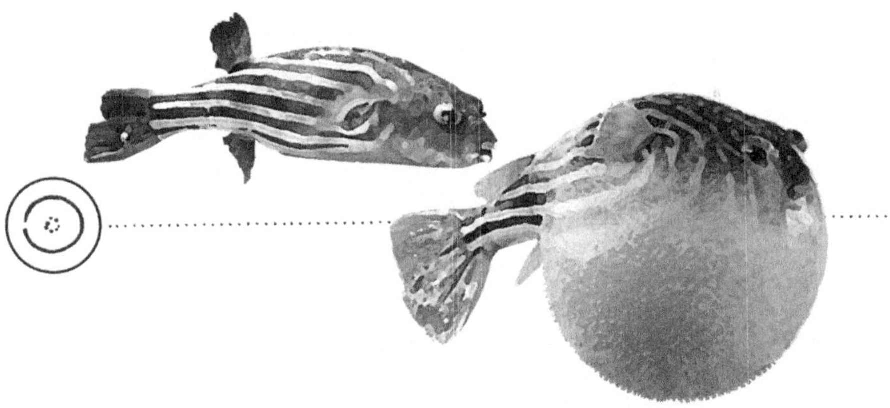

Fig. 34 Expansion. Pufferfishes can inflate their elastic stomach as a defense mechanism

7 Connection Patterns

The following 'family trees' of connections represent a tool to read and understand the complexity and the interrelations between the studied parameters. A complete classification of all systems in Nature would be useful, but extremely difficult, if not impossible to realize. Nature continuously evolves new systems, and a great number of these are most probably still unknown to us. Therefore, in this context, classifications of recurring patterns of organisms that are commonly known are suggested.

The systems are classified according to the body structure. As can be noted in the figures, the structures limiting the range of movement geometries the most are the systems using rigid parts, either internal or external (Figs. 35, 36, 37).

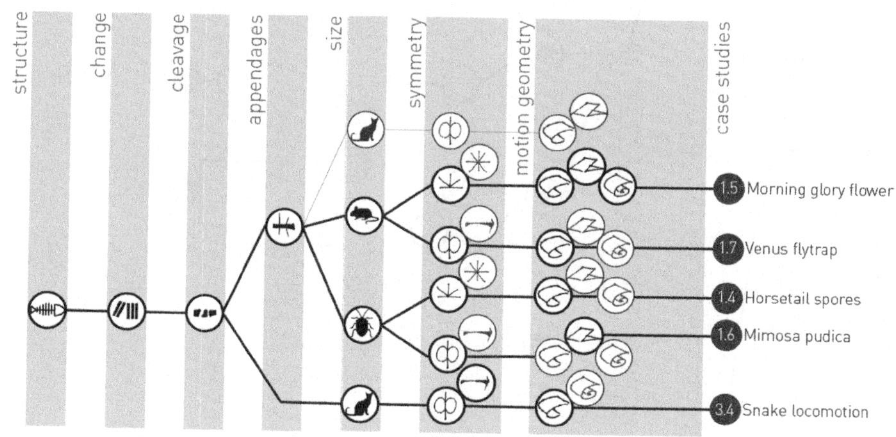

Fig. 35 Geometry of Motion. Connection patterns of endoskeletal body types with the other parameters studied. For case studies, refer to the deepened analysis in Chap. 6

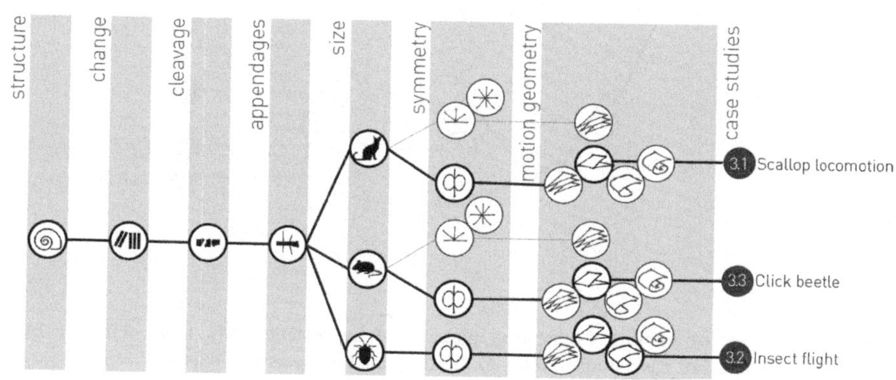

Fig. 36 Geometry of Motion. Connection patterns of exoskeletal body types with the other parameters studied. For case studies, refer to the deepened analysis in Chap. 6

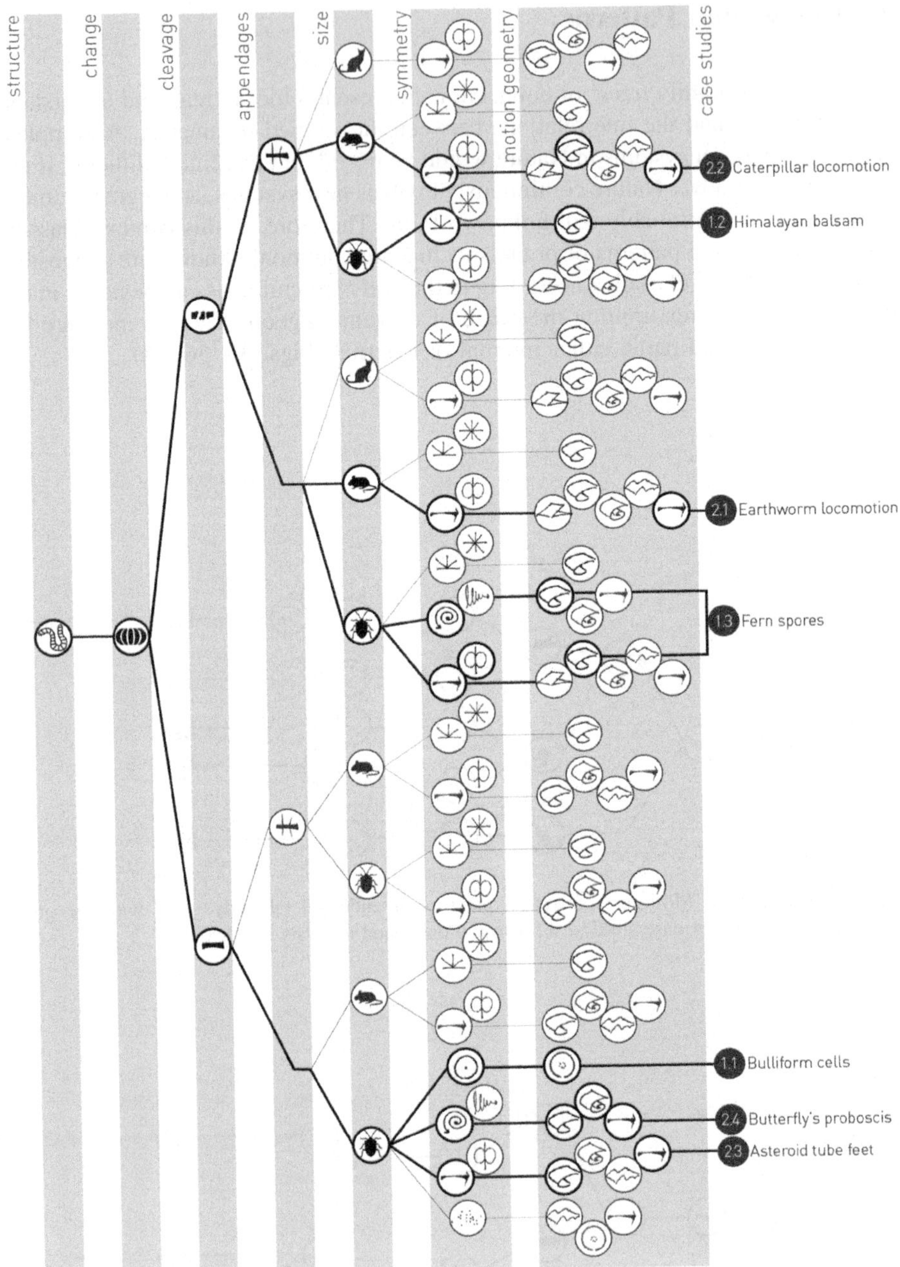

Fig. 37 Geometry of motion. Connection patterns of hydrostatic body types with the other parameters studied. For case studies, refer to the deepened analysis in Chap. 6

References

Arthur W (1997) The origin of animal body plans: a study in evolutionary developmental biology. Cambridge University Press, Cambridge UK

Astley HC (2012) Getting around when you're round: quantitative analysis of the locomotion of the blunt-spined brittle star, *Ophiocoma echinata*. J Exp Biol 215:1923–1929. https://doi.org/10.1242/jeb.068460

Barnes RK (1998) The diversity of living organisms. [electronic resource]. Blackwell Science, Oxford [England]; Malden, MA, USA

Biewener AA (2007) Animal locomotion. Oxford University Press, Oxford, UK. ISBN 978-0-19-850022-3

Bongard JC, Chandana P (2000) Investigating morphological symmetry and locomotive efficiency using virtual embodied evolution. In: Meyer J-A et al (eds) From animals to animats 6, proceedings of the sixth international conference on simulation of adaptive behavior. The MIT Press, Cambridge, UK

Damiens R, Rhee H, Hwang Y, Park SJ, Hammi Y, Lim H, Horstemeyer MF (2012) Compressive behavior of a turtle's shell: experiment, modeling, and simulation. J Mech Behav Biomed Mater 6106–112

Damuth J (2001) Scaling of growth: plants and animals are not so different. Proc Nat Acad Sci USA 98(5):2113–2114

Elder HY, Trueman ER (eds) (1980) Aspects of animal movement. Cambridge University Press, Cambridge, USA

Gruber P (2011) Biomimetics in architecture, architecture of life and buildings. Springer-Verlag, Wien. ISBN 978-3-7091-0331-9

Hirasawa T, Nagashima H, Kuratani S (2013) The endoskeletal origin of the turtle carapace. Nat Commun 4:2107

Kier WR (2012) The diversity of hydrostatic skeletons. J Exp Biol 215:1247–1257, April 15. http://jeb.biologists.org/content/215/8/1247.full

Manuel M (2009) Evolution/Évolution: early evolution of symmetry and pola rity in metazoan body plans. Comptes Rendus - Biologies, n.332 (La theorie de Darwin revisitee par la biologie d'aujourd'hui/Darwin's theory revisited by today's biology), pp 184–209

Moloney J (2011) Designing kinetics for architectural facades, state change. Routledge, New York

Pfeifer R (2000) On the role of morphology and materials in adaptive behavior. In: Meyer, J-A et al (eds) From animals to animats 6, proceedings of the sixth international conference on simulation of adaptive behavior. The MIT Press, Cambridge, UK

Rosen J (2008) Symmetry rules. How science and nature are founded on symmetry. Springer-Verlag, Berlin Heidelberg

Schmidt-Nielsen K (2004) Scaling, why is animal size so important? Cambridge University Press, Cambridge, UK, (1 ed, 1984). ISBN 0-521-31987-0

Shape-Shifting Frog Discovered in Ecuadorian Andes (2015) Sci Child 52(9):12–13

Thompson D'AW (2006) On growth and form. Cambridge University Press, 1961 (Trans: Albus A). Eichborn AG, Frankfurt am Main

Valentine JW (2004) On the origin of phyla. The University of Chicago Press, Chicago and London. ISBN 0-226-84548-6

Valentine JW (na) Bilateria. Access science. Accessed 29 October 2014 from http://www.accessscience.com/abstract.aspx-?id=802620&referURL=http%3a%2f%2fwww.accessscience.com%2fcontent.aspx%3fid%3d802620

Wilson AM, Van den Bogert AJ, McGuian MP (2000) Optimization of the muscle-tendon unit for economical locomotion in cursorial animals. In: Herzog W (ed) Skeletal muscle mechanics, from mechanisms to function. John Wiley & Sons, Ltd., Chichester, UK, pp 517–545

Wolpert L, Tickle C, et al (2011) Principles of development, 4th ed. Oxford University Press

Chapter 4
Biomechanics of Motion

Abstract The mechanics of motion in Nature are strongly connected to the body features selected, making the most out of the kinetic transmission potential of the structures—whether of an elastic or plastic nature. In parallel to the previous analysis of possible geometrical conformations, this chapter categorizes seven main topics, with the relative sub parameters, characterizing the transformation and the efficient transmission of kinetic energy in living organisms.

Biomechanics is a multidisciplinary field, which studies the mechanical aspects of biological organisms and the effects forces have on the motion of the bodies through their structure (anatomy) and function. The topic bridges varied fields of research, from cardiovascular biomechanics, orthopedic biomechanics, sport biomechanics, etc. In many cases, the aim is to understand and optimize the mechanical interaction between parts (Britannica 2016a). In this case, the aim is to break down the parameters defining and characterizing the mechanics of specific types of motion into different categories to simplify the comprehension and the description of the organisms for a future biomimetic transfer of technology.

Control, actuation and transmission are the main topics, which need to find an integrated solution when designing for motion. As specific geometries and types of energy transmission belong to specific types of mechanics, control and actuation categories, the aspects defining the biomechanical parameters of a body can be described as a branching of design possibilities.

In the attempt to present a simple but complete set of highly characterizing parameters, the aspects analyzed belong to a variety of contexts, from the animal to the vegetal kingdom, and combine knowledge from a variety of topics and of disciplines.

S. Persiani, *Biomimetics of Motion*,
https://doi.org/10.1007/978-3-319-93079-4_4

1 Control Center

The control center is, in general terms, the part(s) of the organism, which receive, integrate and interpret incoming information and coordinate response by signaling subordinate parts on how to act. The control center can therefore be described, in all cases, as a processor of information. Like any computing system, the more complex the information it handles, the more energy-intensive the system becomes. The main parameters describing the control center's capacity are:

- Amount of incoming information that can be handled;
- Amount of detail in the information;
- Amount of possible combinations of solutions;
- Response time.

In the case of biological organisms, in many cases the information that needs to be processed to achieve survival is complex and manifold. Not only must animals monitor their surroundings for possible threats or opportunities, but must also follow their own body's needs and mechanics. As energy-efficiency is, at the end of the day, what every organism must achieve to survive, highly energy-consuming processing systems might be useful, but are certainly also very costly. So how to spend less on the energy bill? As we can do in our homes, Nature cuts on energy use and shares available energy as much as possible.

Two different types of control centers can be identified: *decentralized systems* and *centralized systems* can both be found to work independently and on different levels in a same macro system, eventually interacting (Fig. 1).

Decentralization allows cuts on energy use through a low-intelligence network able to answer to basic tasks: specialized minor centers are dislocated to intercept routine information and to deal locally with simple tasks related to it.

Sharing of energy (in terms of information) is achieved as the minor centers are connected to each other through a simple set of rules, which enables them to react as one macro system, minimizing the effort spent on coordination.

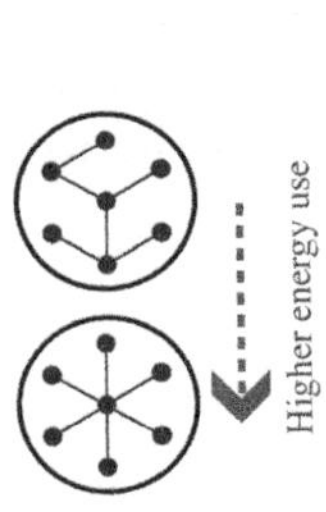

Fig. 1 Types of control center (from the top): decentralized control and centralized control

1.1 Decentralized System

Decentralized systems work within a limited action range and with more simple tasks, but in groups. As clusters of low-capacity processors, consistent amounts of energy can be saved dislocating simple tasks with pre-determined patterns of answers. In other words, decentralizing control centers with specialized tasks (Fig. 2):

- Amount of incoming information is *limited*;
- Amount of detail *can be complex*;
- Amount of possible combinations of solutions are *predictable*;
- Response time is *shorter* than if processed though a central unit dealing with multiple tasks.

Once all tasks that can be independently dealt with are dislocated, space is opened up in the central processor to concentrate on complex tasks that do not have as predictable outcomes.

Example	1. DNA locally deciding though genetic switches over the life and actions of its cell, able to act independently from the central brain system but in unison with its akin 2. Routine body functions as sweating, goose bumps, etc.

Swarm intelligence, as any decentralized system, is based on a population of low-level simple agents interacting locally with one another and with their environment. The overall outcome of these interactions create complex effects, which appear to be working in unison like a single organism. This type of coordination

Fig. 2 Decentralized system: one example of swarm intelligence, a school of fishes swimming tightly in the same direction. Schooling and shoaling are thought to increase both individual security (lower chances of predation, higher chances of foraging) and hydrodynamic capacity (Alexander 2004)

and communication is considered as a form of Artificial Intelligence (A.I.), based on collective behavior in decentralized, self-organized systems (Fox and Kemp 2009).

The rules of interaction between each agent (or system) can be very simple, but, provided that the groups of individuals are sufficiently large, the combinations of interactions can develop emergent behaviors that are very difficult to predict.

- Amount of incoming information is *very limited*;
- Amount of detail is *very limited*;
- Amount of possible combinations of solutions are *limited* for the single individual and *extremely complex* over a population;
- Response time is *short* as choice is limited to imitation of other individuals.

Swarm intelligence is typically a way of controlling large groups of individuals, which need to act as one and decide on recurring problems as which way to move, when to migrate and where to forage. This type of control is therefore unique as it allows control outside the physical boundaries of an organism, and can extend over very large areas and numbers: birds as Dickcissels (sp. *Spiza americana)* have been recorded to be able to flock in several millions of individuals (Basili and Temple 1999).

Example	1. Bird flocks creating changing patterns in the sky; 2. Fishes schooling; 3. Bee swarm

Typical main parameters used to describe fish shoals are:

- Size, given by the number of individuals (often in thousands or millions) and the area covered (often over kilometers);
- Density, given by the number of fish divided by the volume occupied—although density is not necessarily constant throughout the group;
- Polarity, describing the extent to which the individuals are pointing to the same direction.

Many group decisions seem to be taken through *quorum sensing*, an auto-inductive decision-making process based on the coordination and monitoring of changes sensed in the population density, used in many decentralized systems as bacteria (Turovskiy et al. 2007). Quorum response in groups of animals is achieved as the probability of group members performing a given action steeply increases, once a minimum number of individuals in the group already perform that action (Ward et al. 2012).

1.2 Centralized System

Centralized systems generally have a more varied action range, compared to decentralized systems, being able to coordinate subordinate structures of multiple nature. The system requires a central processor, or an element upon which initiation of a command depends, and a communication network for the processor to reach the subordinate agents (Fig. 3).

Example	1. Brain system with a neural network

The processor, or 'brain element', is responsible for processing incoming information and for transmitting the actuation signal. As noted beforehand, the more the complexity and amount of the information processed, the more energy-intensive the brain element. A centralized system working in parallel to a decentralized system, which operates routine information and reaction, allows to optimize its energy and focus on specific tasks:

- Amount of incoming information is *selected depending on the task*;
- Amount of detail is only processed *if necessary*;
- Amount of possible combinations of solutions are *reduced*;
- Response time is *adapted*, depending on the urgency of the task.

The communication network allows the actuating element to transmit the actuation order. Actuating one or few elements connected by an intentional pattern animates a wider area. The design can be very simple and relatively inexpensive as the reaction of one element causes reactions in the successive stages of the chain, resulting in a self-amplification reaching the boundaries of the pattern. Speed depends on the geometry of the pattern and the time lapse between the action of the "brain" and the reaction of the last unit in the chain. This kind of system is easily controllable but relies heavily on a few components: if these are put out of operation, the whole system is severely affected.

Example	1. Chain reaction, or domino effects

Fig. 3 Centralized system: the domino-effect, example of a chain reaction

2 Actuator

The actuator is the "motor" converting energy into motion in the organism. Actuators are in this context classified according to the *source* of the energy triggering the movement, and to the organisms' *processing* of the reaction.

Passive actuation is achieved when forces acting on the organism trigger a reaction to which the organism responds working in line with the triggering factor (> >). The result is a pre-determined outcome of a succession of domino-reactions (see system control center and centralized systems). This involves that at a given signal, the organism will react in a predictable manner and cannot choose to react otherwise.

Active actuation is achieved when the organism consciously intervenes, steering the reaction, also by contrasting the acting forces (> <) to overcome a disturbing factor (Fig. 4).

A further distinction within these two categories of motion actuation defines the amount of energy needed in the actuation process.

Low-energy solutions vehicle efficiently already existing forms of energy that are external to the body. These are mostly latent environmental forces under the form of kinetic energy (wind, rain, other organisms moving in proximity), potential energy (gravity, humidity) or radiation (light, heat). Those low-energy solutions are here called "external".

- *Passive external actuation* is hence the most elementary form of actuation, requiring (virtually) no other energy input than the energy received by the organism from its environment. It is typically used by plants and can be found in many nastic and tropic reactions (see Chap. 2, part 7);
- *Active external actuation* is more complex as it generally requires a sensor-brain system and is therefore mostly used by animals. Typical examples are birds exploiting winds and rising hot air currents to move over wide territories.

Low energy however also means lower security as external actuation is not possible if the environmental conditions do not allow for it. This means that most organisms have developed a long resistance in time, waiting for the conditions to become more favorable, or alternative (active) actuation solutions to compensate for unavailable external energy.

High-energy actuating solutions involve that the organism provides for the actuating energy by itself, raising its metabolic rate. Those low-energy solutions are here called "internal".

Fig. 4 Two main types of actuation (from the left): passive actuation and active actuation

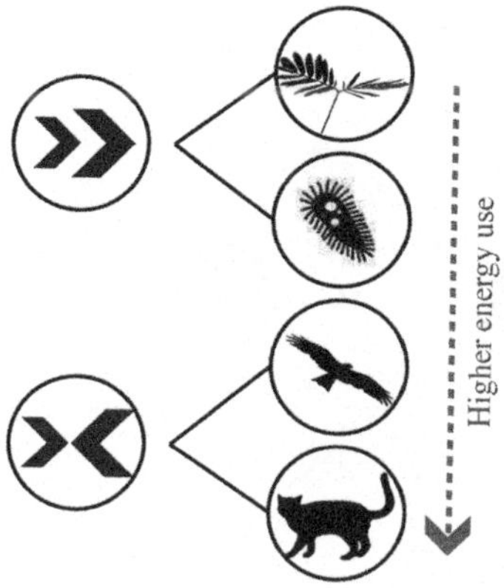

Fig. 5 Types of actuation, subdivided into low and high energy solutions (from the top): Passive external actuation; Passive internal actuation; Active external actuation; Active internal actuation

- *Passive internal actuation* typically does not need a complex network to function, as all actions are pre-determined once energy is provided. This type of actuation can be found in plants, simple mono- and multi-cellular systems, and in the visceral functions of higher animals.
- *Active internal actuation* is by far the most complex and energy-consuming type of actuation. It requires the organism to provide the energy to power action and a complex brain-sensor system to control it (which is extremely energy consuming). Therefore, this type of actuation can be found in animals only.

High-energy solutions are typically used for functions requiring a high degree of control (breathing, growth, reproduction, escape, etc.) of the timing and the manner to release energy (Fig. 5).

2.1 Passive External Actuation

Passive external actuation is broadly used in systems lacking a central nervous system: the organism fully exploits its environment's changing physical conditions to move, relying on pre-determined chain reactions of chemical and mechanical nature. The movement is low-energy and in some cases mechanical work achieved is a direct consequence of the triggering factor's energetic input itself. In most cases, to a specific trigger corresponds one reaction only. Action and reaction have a 1:1 relationship in real time as input/response/outcome: to trigger a new adjustment, the input must change (Fox and Kemp 2009) (Fig. 6).

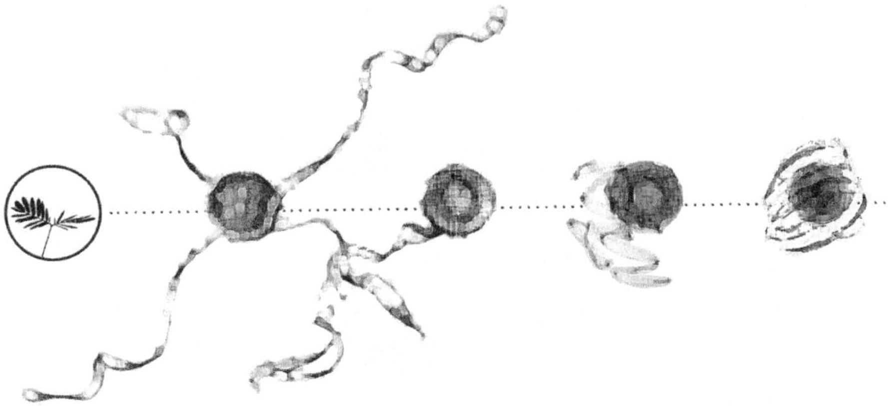

Fig. 6 Passive external actuation: horsetail spores (so. *Equisetum arvense*) have four ribbon-like appendages (elaters) which can deploy and re-wrap around the spore, depending on the weather conditions. This behavior facilitates dispersal, as the spores can wander around with open elaters while the weather conditions are dry (Marmottant et al. 2013). The main source of energy is external (humidity) and the organism uses the energy input to passively adapt

Example	1. Grass leaves curling in wet weather conditions; 2. Horsetail spores (sp. *Equisteum arvense*) 3. Venus flytrap (*sp. D. muscipula*)

Being the most elementary form of actuation, requiring (virtually) no other energy input than the energy received by the organism from its environment, passive external actuation is foremost used by plants for basic functions as seed dispersal, which have functions that can be directly associated with the most predictable environmental conditions.

2.2 Passive Internal Actuation

In *passive internal actuation,* the body of the organism itself is in charge for the action, triggering a series of chain-reactions to produce specific movements. This kind of actuation is broadly used to control pre-determined basic movements necessary for the good functioning of the organism. The predictability of the action-reaction on one hand ensures security in the execution and the correct operation of functions that are vital to the organism. On the other hand, it leaves space and energy for more complex functions (requiring *active internal actuation*) to be consciously controlled. Passive internal actuation can broadly be found in visceral functions of higher organisms (Fig. 7).

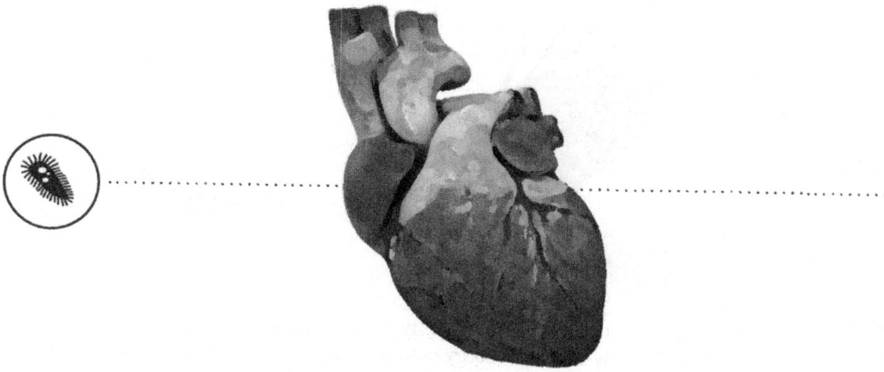

Fig. 7 Passive internal actuation: example of a human heart. The body itself produces the energy for the pulsations, which are controlled through pre-determined movement patterns and eludes conscious control. The relatively simple pumping movements change predictably together with other chain reactions in the body: changing gait from walking to running increases the beating frequency

Example	1. Heart-blood-vessel system 2. Eukaryotes and procaryotes

It is interesting to note that some functions, as breathing, belong to those visceral functions that are, generally, autonomously controlled by pre-determined patterns: we don't need to learn how to breathe, and we don't need to remember to breathe. However, unlike for the heart beating, these specific patterns can be overridden upon need. An obvious reason is, in our case (sp. *Homo sapiens*), that we are able to change medium: as we swim underwater, our lungs cannot by themselves know when to hold the breath, exhale and inhale again, so the brain 'overrides' the standard bodily breathing orders and intervenes changing the breathing pattern (though active internal actuation).

2.3 Active External Actuation

Active external actuation is broadly used by organisms with a central nervous system for movements with a minimum of effort: it exploits the forces present in the medium surrounding it to move, using very little energy to control and steer the movement into the wished output. The great difference with the passive external

Fig. 8 Active external actuation: soaring birds typically exploit the kinetic energy in their surroundings to move efficiently at a low energy output: rising hot air currents are used to rise and glide over longer distances. The birds use their own energy only for small steering and adjusting movements: the main source of energy in the displacement is external, and the organism actively controls it

actuation is that the organism itself decides when to break a given equilibrium, changing the movement pattern into another one by producing a short burst of counter-work (through active internal actuation). In other words, one triggering action potentially corresponds to a multitude of reactions: the organism chooses the outcome; depending on the movement it wishes to achieve (Fig. 8).

Example	1. Gliding birds; 2. Soaring birds; 3. Fishes making use of the buoyancy effect

2.4 Active Internal Actuation

Active internal actuation requires a complex and developed sensor-brain network to interpret the exterior conditions, measure, decide and react. To this type of actuation belong most behavioral movements of animals as hunting, nesting, social interaction, learning and so on. The organism itself provides the energy and work necessary to obtain the wished motion, and hence requires higher collection of energy (through feeding) and energy reserves (chemically breaking down and transforming the energy in the food). This type of actuation has a much higher energetic cost compared to the others, but also allows fast and varied responses to ideally any input, which can compensate providing huge evolutionary advantages (Fig. 9).

Example	1. Mammals; 2. Reptiles; 3. Birds

Fig. 9 Active internal actuation: a fox hunting in the snow. The fox stalks its prey at a distance, then jumps into the air and dives into the snow, pinning the moving target between its paws. Generally, hunting requires an extremely complex coordination between the identification of the target and prediction of its movements, then physical coordination for the attack. Such complex movements require an advanced decision-making and control system enabling reaction to unpredictable situations

3 Agonist Mechanisms

A movement can be divided into two main phases: movement and counter-movement. *Agonist movement* is referred to as the action, which brings the organism out of its initial state of equilibrium. The counter-movement, or *Antagonist movement* is responsible for bringing the mechanism back to the initial position.

Agonist mechanisms deal with the mechanics of motion and more precisely, with the body structure that initiates the movement. The most simple and immediate types of agonist mechanisms use changes in the environment to actuate motion. Muscles on the other hand are more complex agonist mechanisms, requiring a nervous network, hence are usually used by complex organisms (Fig. 10).

Hygroscopic mechanisms and *Pressurized compartments* react to the change of the conditions in the environment (moisture, heat, etc.). These types of mechanisms have generally long reaction times as energy is built up over longer periods of time, loading mechanisms that then release all the energy at once. These mechanisms are therefore quite common in plants that use them for seed dispersal, while few examples can be found in animals. These two types of antagonist mechanisms should not be confused with the mechanics of hydrostatic systems used by soft-bodied animals to locomote. In this latter case, compartment systems are used as amplifiers of an agonist muscle force to gain mechanical advantage.

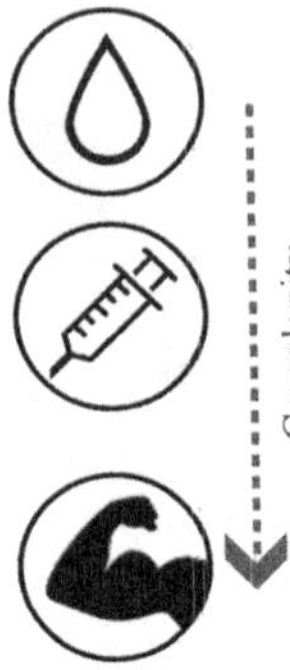

Fig. 10 Agonist mechanisms (from the top): hygroscopic, pressurized compartment, muscular

Muscular mechanisms are the most common agonist mechanisms in the animal kingdom. Generally the force produced by the organism is due to the muscle itself only in a small part, as force is amplified though a great number of means. The muscle can also act as a trigger to release energy gathered though other means.

The choice and combination of mechanisms to actuate the energy and to vehicle it effectively at a low energy cost strongly depend on the physical and mechanical proprieties of the connected body parts (Nachtigall 2005).

3.1 Hygroscopic Mechanisms

Hygroscopic mechanisms absorb water to expand and contract their structure as the environmental conditions change and the water evaporates. This way, movement is induced through reaction to moisture fluctuations (Fig. 11).

Example	1. Geranium palustre (sp. *Geranium palustre*)

Fig. 11 Hygroscopic mechanism in the Geranium palustre: elastic energy is released by a hygroscopic mechanism as the pod shells dry up

Geranium palustre has an explosive-hygroscopic mechanism: when dry, its flat pea shells tear apart and curl up instantly throwing off the seeds. See detailed case study in Chap. 6.

Example	2. Equisetum arvense's spores (sp. *Equisetum arvense*)

Equisetum arvense spores have long, ribbon-like appendages, which curl in humid weather. As moisture fluctuations are recurrent in these plants' habitat, the bands intertwine with each other, pushing the spores away from the plant, protecting the spores and allowing dispersal by increasing aerodynamic drag (Marmottant et al. 2013). See detailed case study in Chap. 6.

3.2 Pressurized Compartments

Single pressurized compartments work with a single pocket or capsule in which pressure can be changed. In many of the cases studied, the compartment is set under an increasing pressure, until it eventually bursts, releasing kinetic energy in an explosive-like manner (Fig. 12).

Example	1. Dung cannon (sp. *Pilobolus crystallinus*) Shooting distance 0–2.4 m, initial speed 10 m/s. This fungus builds up a high turgor pressure to shoot away its spore-capsule 2. Exploding cucumber (sp. *Ecballium elaterium*) Shooting distance 0–10 m, inside pressure 0–5 bar

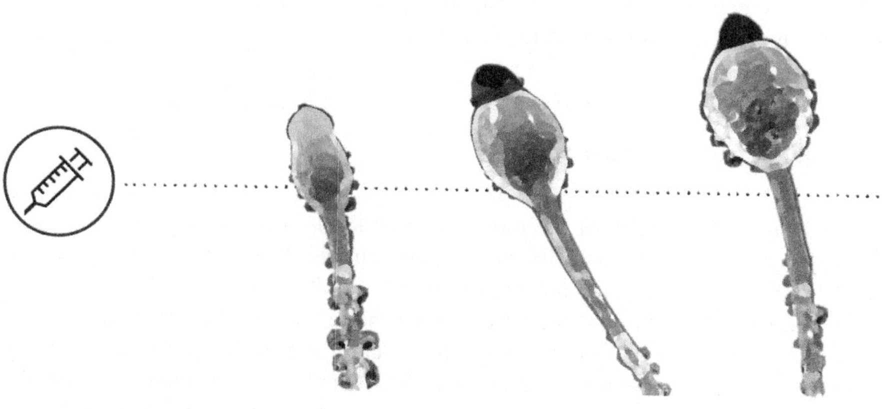

Fig. 12 Pressurized compartments: three steps of a Dung cannon spore building up internal pressure before shooting away the spore capsule when the pressure overcomes the resistance of the linkage keeping it in place

The fruits of the exploding cucumber hang loosely upside down on curved stalks. The inner compartment containing seeds is considerably over-pressurized, tensing the elastic skin: at the slightest touch the fruit hooks off the stalk throwing the seeds away.

Example	3. Hydras (gen. Hydra) Shooting speed 30 m/s, max speed reached in 4000 sec-1 harpoons their preys with high pressurized capsules located on their tentacles

Combined pressurized compartments work as a series of juxtaposed pockets or capsules with changing interior pressure. Compared to the single-compartment system, combined compartments typically achieve more complex moments, depending on the special organization of the compartments and the bonds between compartments.

Example	1. Himalayan balsam (sp. Impatiens glandulifera) Shooting distance 1–2 m, inside pressure 0–20 bar

In the Himalayan balsam, five over-pressurized cushion-like shells held together by a light interstitial tissue and an inner layer of longitudinal fiber cells compose the pod containing the seeds. When mature, the tissue and the fiber cells dissolve triggering the explosion mechanism. The pod explodes and rolls up instantaneously when touched (Nachtigall 2005). See also the deepened case study in Chap. 6.

This type of actuation is quite common in plants as it allows them to prepare under a longer time and burst when the environmental conditions are the most favorable, releasing all energy in one single action (mostly to throw their seeds as far away as possible). A one-time action however means that the agonist movement is most of the time not coupled to an antagonist movement and can therefore not be repeated once the energy has been released.

On the other hand, in the case of animals using pressurized compartments, an antagonist mechanism is most of the time present allowing a cyclical reversal between agonist and antagonist mechanism.

3.3 Muscular Actuation

Muscular actuation is one of the most common kinds of actuation mechanism and can be found in a great majority of moving animals, from vertebrates to invertebrates and insects. Muscles are made out of a soft tissue of cells, which upon stimulation contract simultaneously changing the length and the shape of the cells. Muscle action can be involuntary, as for the cardiac and the smooth muscles, or voluntary as in skeletal (or striated) muscles, which are consciously controlled. Skeletal muscles in particular are stimulated by electrical impulses transmitted from the brain through the motor nerves (motoneurons), but also react to reflexive nerve stimuli, which don't pass through the brain (as the withdrawal reflex upon touching something hot, sneezing, etc.) (Fig. 13).

Fig. 13 Muscular actuation in a fish: sections of a salmon's muscle bundles. The fish has identical sets of muscle mirrored along the sagittal plane which allows it to alternate action on left and right side, achieving a carrangiform swimming motion

The potential of a muscle to do work is directly related to its volume. Force depends on activation, length and speed of the muscle's contraction: fibers arranged parallel to the muscle's longitudinal axe (muscle A) produce twice the force as a muscle of the same volume with fibers arranged inclined to its axe (muscle B). Muscle B will however have twice as many fibers and be twice as strong as muscle A (Stein et al. 2000). As muscles mostly make use of rigid or deformable bodies to amplify the force, the amount of force applied to the external environment finally depends on the mechanics involved in the system.

Example	1. Skeletal muscles in vertebrates 2. Insects

Insects can typically perform much better than other animals compared to their small body size. The ant species *Formica exsectoides* has been theorized to be able to lift up to 5.000 times its own body weight (Nguyen et al. 2014) and froghoppers (fam. *Cercopidae*) can jump up to 115 times their body length (Burrows and Sutton 2008). This is because the energy output is much higher in relation to their body mass: about 400 times in the case of the froghopper.

Insects have also evolved very particular types of mechanics for flight. Insects with very high wing stroke frequency have evolved *asynchronous muscles* that allow more frequent contractions than the rate at which the nerve impulses reach the muscles (see case study on Insect flight in Chap. 6), allowing faster wing beats.

4 Antagonist Mechanisms

Because most mechanisms cannot actively produce both movement and counter-movement, force must be transmitted so as to return the mechanism to its initial state (Kier 2012; Biewener 2007; Janocha 1992). Hence, agonist and antagonist mechanisms work in pairs (Fig. 14).

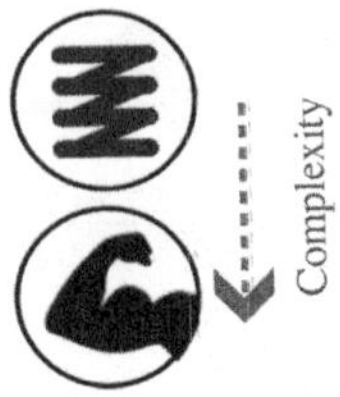

Fig. 14 Antagonist mechanisms (from the top): spring, incompressible fluid and muscular

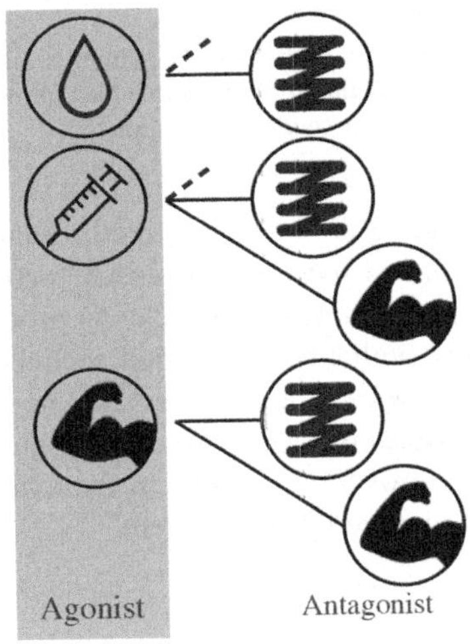

Fig. 15 Agonist-antagonist combinations. The dotted line indicates the existence of mechanisms not coupled to any antagonist (as in explosive, one-shot mechanisms)

Spring elements and *pressurized compartments* react as antagonists as they regain their state of equilibrium, which the agonist had countered by producing work in order to move.

Muscular antagonism, as muscular agonism, needs actuation though a nervous system, and requires therefore a more complex body structure and organization (Fig. 15).

4.1 Spring Antagonism

Spring antagonism is an extremely common and simple type of antagonism. The spring element, which can be any kind of flexible material or structure, absorbs energy from the agonist movement and releases it to return the body to its initial position. Spring antagonism is therefore a very versatile type of mechanism and is therefore very diffused. It can be found in combination with most agonist mechanisms, both in plants and in animals (Fig. 16).

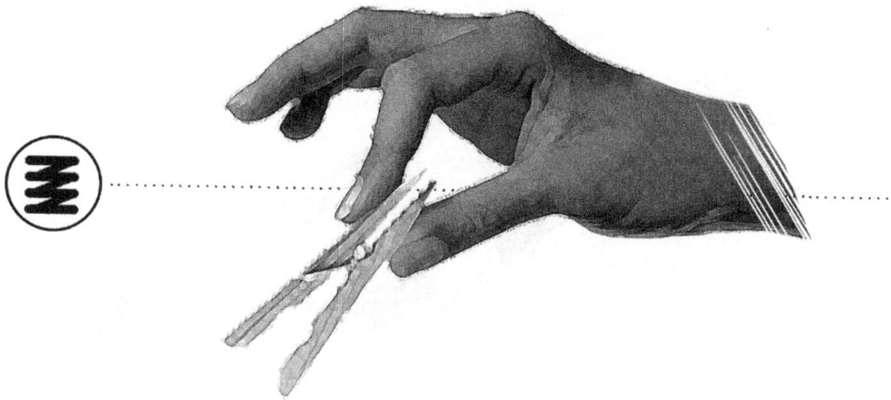

Fig. 16 Spring antagonism. In the case of clothes pegs, the closing action is achieved through a spring element

Example	1. Clam's shell opening mechanism 2. Pistol shrimp (fam. Alpheidae), which uses its oversized "snapper" claw as an acoustic weapon to stun or kill prey animals

The pistol shrimp has evolved an underwater acoustic weapon in one of its claws to stun its prey. The mechanism is similar to a pistol, made of two parts: a joint allows a "hammer" to be locked backwards and is then released, snapping into the other part of the claw. The sound produced does not come from the two claw halves hitting each other, but from the implosion of a cavitation-bubble formed in a fast water jet, squeezed out from between the claws (Versluis et al. 2000). The cavitation bubble generates acoustic pressures up to 80 kPa at a distance of 4 cm from the claw. As it extends out from the claw, the bubble reaches speeds of 97 km/h and releases a sound reaching 218 dB (Heberholz and Schmitz 1999). In this case, the agonist mechanism is achieved by a muscle drawing the hammer backwards (loading the spring mechanism), and the snapping itself is due to the spring mechanism releasing the loaded energy.

Example	2. Earthworm (sp. *Lumbricus terrestris*)

Soft bodied systems, like the earthworm, amplify their movements by manipulating a flexible body containing a liquid, which by its incompressible physical proprieties deforms the body to maintain the same volume when put under pressure. The flexible walls of the body can in part act as spring antagonism, returning the body to the initial shape once the agonist force is removed.

4.2 Muscular Antagonism

Muscular antagonism is broadly used by most animals. It is found in vertebrate systems where the action of a muscle is amplified by actuating rigid elements, and

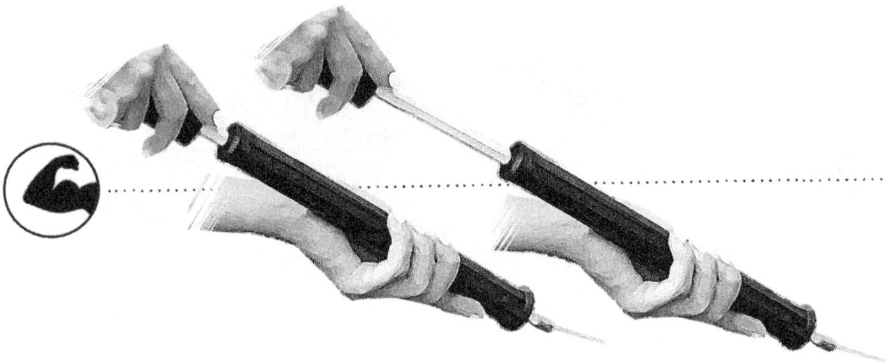

Fig. 17 Muscular antagonism. In this example, to use a bicycle pump, both pumping movements (pushing and pulling) must be made using muscle force

in invertebrate systems where the muscle force is amplified by the deformation of a flexible body containing a liquid.

In some cases, agonist and antagonist are activated simultaneously to obtain stability by stiffening the joints (Le et al. 2017). Muscular agonism and antagonism is a very common combination and lies at the base of motion in all vertebrates as it allows more precise movements and control (Fig. 17).

Example	1. Human arm 2. Tongues

In complex muscular systems, the distinction agonist/antagonist can become more difficult as a multitude of effects and actions can be achieved by combining contraction in one or more muscles. In composite muscles as the tongue, various components like longitudinal, transverse, horizontal muscles enable very precise and complex movements. This feature can also be seen by the variety of the functions that vertebrates have evolved in their tongues, from the cat cleaning its fur to the prehensile tongue of the chameleon catching prey.

5 Power Amplification

Power amplification allows the transmission of motion, by arranging the biomechanical body parts to transmit force from one organ to another, in order to create mechanical advantage (Nigg et al. 2000; Shigley and Uicker 2010). The initial spark of energy is multiplied by many times, and is used both in agonist and antagonist mechanisms (Fig. 18).

Elastic amplification makes use of the flexible proprieties of selected body parts to store mechanical energy by transforming kinetic energy into potential energy and reversely, in a continuous cycle of energy exchange. Ideally, these cycles require

Fig. 18 Power amplification (from the top): elastic, deformation of a volume, indirect actuation, chemical reaction

only an initial input of force, and then goes on switching from one state to another alternatively (Viegas 2005; Britannica 2016b, c).

Deformation of a constant volume is typical of soft-bodied structures with a fluid within flexible compartments.

Indirect actuation is mostly used in rigid or stiff bodies exploiting lever effects of different complexity.

Chemical reactions are at the base of the energy exchange within any living organism, however the deliberate ignition of a reaction to achieve motion seem to be used in the technical world exclusively, and is therefore not further analyzed.

Power amplification mechanisms often work together, in a varied multitude of combinations in the same organism. As an example, the Geometridae caterpillars (fam. *Geometridae*) combine muscular agonism and antagonism with a soft-bodied pressurized compartment and elastic body shell to achieve an inching locomotion, which is based on a simple lever system though direct actuation.

5.1 *Elastic Amplification*

Elastic structures in the body function mostly as energy storage and power amplifiers, more than to truly initiate action. An elastic element or material will store potential energy as the result of its stretching or compressing, which requires previous action upon it. When a spring is not stretched or compressed, there is no elastic potential energy stored in it, and it is said to be at its equilibrium position. From this resting position, the spring needs to be loaded with an input of kinetic Energy (through compression or tension). When the Force acting on the spring is then removed, the spring will return the potential Energy in the opposite direction to which it has been loaded, restoring its initial position of equilibrium (Viegas 2005; Britannica 2016d) (Fig. 19).

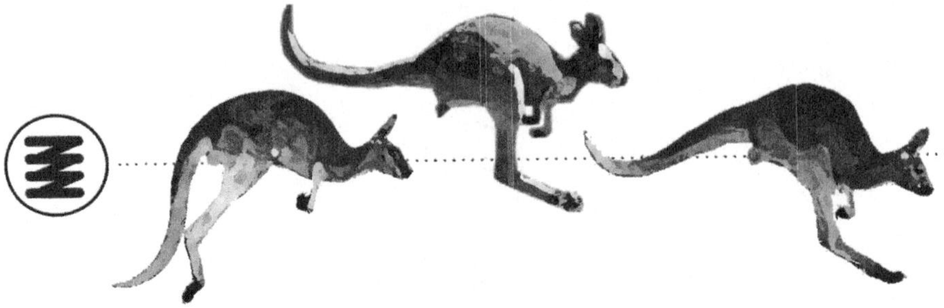

Fig. 19 Kangaroos make good use of the natural and geometrical elasticity in their legs and body, bouncing forward (and swinging their long tail) using virtually no other energy than the potential gravitational energy collected in the first jump and returned by the spring effect

In this sense, the elastic energy is more significant as contributor to the actuating efficiency than as actuating force. When stored energy is released very rapidly in the final stages before an action, it improves the action's performance by the principle of a catapult: the energy returned in the recoil equals to the work done to stretch the spring, but is returned at a greatly increased rate.

Natural elasticity refers to the flexible and elastic proprieties possessed by one or more of the substances a system is made of. These features are strategically located to provide easy recovery of energy, combined with control of the movement when released.

Example	1. Tendons. Tendons have greater elastic recoil and less heat dissipation than a muscle, providing as well good tensile strength (Alexander 2000). When running, energy is stored in tendons and ligaments at each footstep and returned at takeoff (Nigg et al. 2000) 2. Clams and bivalves' ligaments

Clams and bivalves have an unusual opening/closing system, worked by only one muscle: contraction closes the shell, compressing a flexible hinge ligament between the shells. The opening is automatically regulated by the tensile antagonism of the linkage, so when the contracting muscle relaxes the ligament recoils elastically opening up the shell. Most other skeletal systems need muscles in antagonistic pairs to work the opposite effect (Britannica 2012). See also detailed study on clam's locomotion in Chap. 6.

Example	3. Muscle fibers

Muscle fibers also have elastic proprieties, but less than tendons. An isometrically contracting muscle fiber, suddenly allowed to shorten will function exactly like a rubber band. However if the force is not rapidly released, the muscle recovers adapting to its new length unloading the spring-effect (Alexander 2000).

Example	4. Elastic Skin, Connective tissues

Fishes seem to transmit part of the forces along their body through their skin acting as a structural exotendon (external tendon). Sharks have a cross-fibred array of collagen fibers beneath the skin, which are used as an external body tendon transmitting forces with a great mechanical advantage (Herzog 2000).

Surface tension. Mechanisms using surface tension need a long preparation time during which a physical process progressively puts a surface into tension, as when loading a spring. When the constraining factor is removed, the system finds itself drastically imbalanced and snaps back to the original position.

Example	1. Venus flytrap (sp. *Dinonaea muscipula*) is a carnivorous plant capable of movement rapid enough to trap its preys

The Venus flytrap's leaves present a rigid curved border and a slightly bulging leaf surface. The changing turgor pressure inside the leaf tenses the convex surface bringing the rigid border to its resistance limit, as close as possible to a flat surface. When an insect lands on the leaf and triggers the tiny hairs on its surface, resistance drops, geometrical balance is broken and the convex surface instantly deforms into a concave surface popping the leaves closed around the prey. See also the deepened study on the Venus flytrap in Chap. 6.

Example	2. Fern (sp. *Dryopteris filix-mas*)

A double-layered "spine" surrounds the Fern's spore capsules: a thin outer protecting screen and an inner thick comb-shaped tissue, which puts the capsule under high elastic tension. The capsules are held together from the inside by the strong cohesion of immature spores. When the spores ripen and dry, cohesion drops and the elastic spine pops the capsules up spreading out the spores. See also the deepened study on the Fern spores in Chap. 6.

Geometric elasticity. Elasticity can be created in semi-rigid elements (or systems made of semi-rigid elements) to store and release, or dissipate energy, through deforming surfaces and structures. These parts are typically built in circular, helical, curved and bent V- and Z-geometries (Vincent 2008; Biewener 2007; Thompson 2006; Alexander 2000; Nigg et al. 2000).

Example	1. Leg elements based on V- and Z-shapes: camels, kangaroos, horses, etc.

Kinetic and gravitational potential energy are strongly interdependent throughout the movement. Force plate records show how the force exerted by the body when it touches the ground is more or less aligned with the leg. The foot is first set down in front of the trunk and the force slopes in such a way to decelerate the falling body; its kinetic energy and gravitational potential is at the minimum as the trunk passes over the supporting foot; later in the step the foot is behind the trunk and the force slopes in such a way as to accelerate and rise the body. The deformation of the shape of the leg acts as a spring element, absorbing and returning the kinetic energy.

Example	2. Curved and bent shapes: Human bones 3. Elongated shapes: Grass stem, Cane stalk, Bird feathers are flexible and allow the wing to bear without stalling thanks to an elastic behavior

5.2 *Deformation of a Constant Volume*

Deformation of a constant volume (generally a fluid) is widely used in most organisms to vehicle movement, from plants to animals. The fluid being incompressible, contraction in one dimension results in extension in another dimension and is hence a very simple way to effectively amplify a force (Valentine 2004) (Fig. 20).

Example	1. Soft-bodied animals 2. Starfish's tube feet

Fig. 20 An example of an incompressible fluid amplification: the tube feet of a starfish are connected to a "pouch" filled with water. When the pouch is squeezed, the water flows into the tube foot, which elongates. See also deepened case study on starfish locomotion in Chap. 6

5.3 Indirect Actuation

Indirect actuation is typically used in organisms built with rigid elements relying on the use of advanced lever systems. To understand the difference with direct actuation, two examples are compared (Fig. 21).

In *simple indirect actuation*, movement is achieved by operating a basic lever-pivot system. The system's power is proportional to the direct force of the muscle (Nachtigall 2005) and the frequency of the effort is limited (0–50 Hz).

Example	3. Flying muscles in most birds

Fig. 21 An example of force amplification (from the top): a swimming athlete and a rowing athlete. In the first case, the lever system (the arms) is simpler than in the second case (arms + oars). If the muscular agonist force is the same in both cases, the force exerted at the end of the oar will be much higher, compared to the force exerted by the swimming athlete at the end of his hands

Fig. 22 Chemical reaction. No organism in nature has been identified to amplify its movements through the use of a chemical reaction

Example	4. Fast flying insects: flies (or. *Diptera*), bees (gen. *Apis*) 200–250 Hz

In *advanced indirect actuation*, the lever-pivot system is more complex and allows higher speed and force than the direct muscle's capacity. This type of structure can be found in body parts oscillating at high frequencies (0–1.000 Hz). When working at very high frequencies, as in insect wings, the duration of the effort is shorter as it produces heat, which can damage the structure. High frequencies are therefore generally used over small distances (Nachtigall 2005).

5.4 Chemical Reaction

Movement triggered by chemical reaction is broadly used in mechanical engines powered by the ignition of fossil fuels. In nature, although all life and movement can in the end be broken down to chemical transformations, the organisms using chemistry to directly actuate motion are, if known, very few and will not be further analyzed (Fig. 22).

Example	1. Bombardier beetle (sp. *Brachinus crepitans*) combines chemical reactants in a violent exothermic chemical reaction expelling a noxious gas-product as defense (Piper 2007; Isaak 2003). The chemical reaction is not known to have effect on the kinetics of the animal

6 Mechanics of Hydrostatic Systems

In hydrostatic systems, mechanical amplification is not transmitted through rigid skeletal elements and muscle contraction, but by deforming geometrically the body through the displacement of an incompressible fluid. Scaling and deforming the elastic envelope can control a various array of movements and change in stiffness. Moreover, elasticity means reversibility of the movement in defined ranges of stress (Fig. 23).

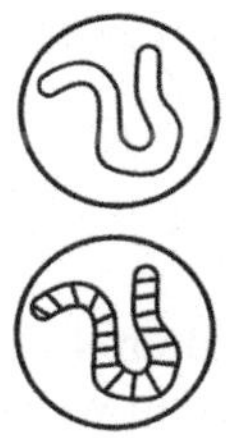

Fig. 23 Two types of hydrostatic system's mechanics (from the left): Single compartment systems and multiple compartment systems

The mechanics work differently whether the compartments are single, multiple and juxtaposed or multiple and connected. Power amplification is usually the combined effect of a deformation of the compartment(s) and the elastic flexibility of the compartments' walls.

These systems are of primary use in plant's cells and in soft-bodied organisms, however variations of the same principle is recurrent even in vertebrate organisms as actuation of 'secondary body parts' as tongues, proboscises and penises (Kier 2012; Nachtigall 2005; Valentine 2004). Although compartment systems can be found in most organisms, whether as body structure or for internal functions, single and multiple compartments have not been found to work combined to each other.

6.1 Single Compartment Systems

Single compartment systems use the physical proprieties of an incompressible fluid to modify the shape of a single cavity (Fig. 24). Volume of a cylinder of length [l] and radius [r]:

$$[V = \pi r^2 l]$$

Fig. 24 A water bottle: an example of the functioning of a single compartment system. A flexible container is used to pressurize the fluid it contains. As the water bottle is squeezed, the water (or the air) inside it is expelled at a higher pressure

Assuming constant V, the equation can be written with the radius differentiated with respect to the length:

$$[dr/dl = -r/2l]$$

Single compartment systems are often designed to interact with the surrounding environment, filling the body cavity with water or air from the exterior environment to inflate the compartment. Motion can however also be triggered to some extent by the tightening of muscles and deformation of the flexible compartment, once it has been filled with a fluid.

Example	1. Sea anemone (sub. *Actiniaria*) 2. Squids (Or. *Teuthida*)

Squids have evolved water jet based locomotion, working through the ejection of a bolus of liquid at high velocity from an internal body cavity. An antagonistic mechanism (powered by muscles or elastic antagonism) expands the cavity refilling it with water for the next jet (Biewener 2007).

Juxtaposed compartments are a variation of the single compartment system, where single compartments (usually not directly connected to the exterior environment) are positioned next to each other so that the movement from one compartment can be transmitted to the next. This type of mechanics is at the base of the structure and activation of muscle fibers.

6.2 Multiple Compartment Systems

Multiple compartment systems are hydrostatic skeletal support systems, which use the displacement of fluid from one compartment to another one, changing shape to create movement and to exert force (Fig. 25).

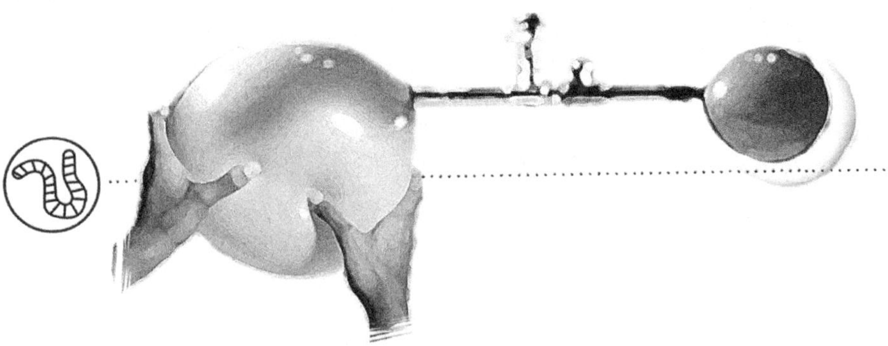

Fig. 25 Mechanics of a multiple compartment system: two balloons are inflated at different diameters and connected by a valve. When the valve is opened, the balloons' size will change slightly but mostly remain of two different sizes, due to the elastic nature of the balloons' surface. When the bigger balloon is squeezed, air is forced into the other balloon, which then grows in size while the former shrinks

Longitudinal stress [L] in the wall of a thin-walled pressured cylinder of radius [r], thickness [t], pressure [p], total force in the longitudinal direction [F] and cross-sectional area of the cylinder wall [A]:

$$[L = F/A = \pi r^2 p/2\pi rt = pr/2t]$$

Circumferential stress [C]:

$$[C = 2pr/2t = 2pr/2t = 2L]$$

Circumferential stress being twice the longitudinal stress in a pressurized cylindrical vessel, we can conclude that to produce equal stress on the body, circumferential muscles should be twice the area of the longitudinal muscles.

Example	1. Pump systems, hearts, vessels and veins

Pump systems are based on the rhythmic contraction of a muscle, pumping a liquid flow through a system. Vertebrate heart systems are very efficient and complicated: composed by one to two receiving chambers (atria) and by one to two discharging chambers (ventricles). Each chamber has its own musculature, which ensures a uniform contraction of all surfaces, to prevent stagnation and to ensure an optimal blood-flow.

Not only for pumping blood, but also for introducing liquids: blood sucking insects often pump saliva into their prey's wound to prevent the blood from coagulating in their suction channel.

Arteries and veins—The vessel system counts about a dozen support pumping-mechanisms the heart cannot make without. The thick arterial walls have an oblique musculature, which can slightly contract the internal cavity.

Example	2. Earthworm (sub. *Lumbricina*)

Earthworms use waves of circumferential muscle contraction in a posterior–anterior direction, causing segmental elongation, and forming retrograde wave (waves moving in the opposite direction to the locomotion direction (Quillin 1999; Chapman 1950; Gray and Lissmann 1938). See also deepened study on the Earthworm in Chap. 6.

7 Mechanics of Rigid Systems

Rigid systems are based on the mechanical proprieties of plastic elements. Rigid transformations do not alter the geometric proprieties of the parts (lengths, angles, shapes in Euclidean geometry). These transformations allow rotation and translation. Plastic elements are usually pre-shaped systems, the geometries are therefore not as flexible as in elastic elements: on the other hand, defining a specific morphology of all constituent parts allows a much easier construction. All constraints

Fig. 26 Mechanics of rigid systems (from the left): levers, inclined planes or wedges; joints; chain systems

and possibilities of the system are defined and embedded in all the possible moving patterns. Most rigid structures in biological organisms are vertebrates, characterized by the segmented subdivision of the internal linear bearing structure of the spine. The subdivision allows rigid transmission of movement only in pre-defined directions. The shape of the vertebrae is adapted to every specific organism to optimize posture, resistance and movement (Fig. 26).

Contrary to the compartment systems typical of hydrostatic bodies, the mechanics of motion transmission in rigid bodies often work combined, in the same or in different parts of the structure to, together, optimize the amplification and control of motion.

7.1 Levers

The *lever* is a movable rigid beam that pivots on a fulcrum attached to a fixed point, amplifying the input force into a greater output force by applying them at different distances from the fulcrum. In an ideal lever no energy is dissipated nor stored, the power into the lever equals the power out. Points farther from the pivot move faster than points closer to the pivot, as power is the product of force and velocity, a force applied to a point farther from the pivot must be less than the force located at a point closer to it (Sclater 2011; Shigley and Uicker 2010; Nachtigall 2005) (Fig. 27).

Fulcrum in the middle. The effort and load are positioned at opposite ends, as in a playground seesaw or a pair of scissors (Sclater 2011).

Example	1. Meadow sage (sp. *Salvia pratensis*)

The meadow sage has a round plate located below the pivot point: when a bee presses it in search of nectar, the curved stylus bends and powders the bee's back with pollen (Nachtigall 2005).

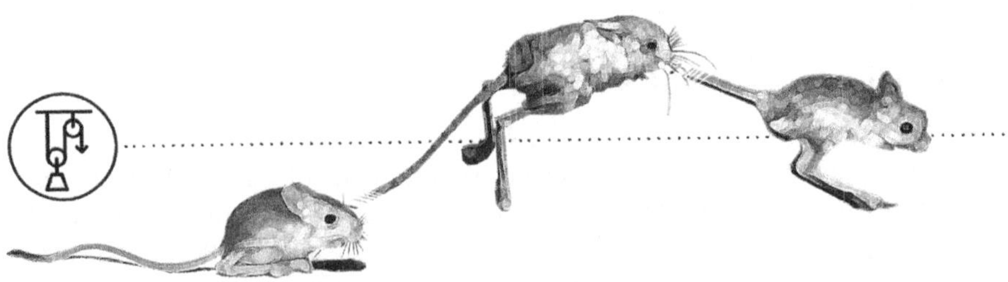

Fig. 27 The lever system in the strong hind limbs of a 5–15 cm Jerboa (fam. *Dipododae*) allows this hopping rodent to jump as far as 3 m in a single jump (Britannica 2018)

Example	2. Crabs' chelae, Birds' beaks

Scissor structures in nature are mostly made out of osseous members which can be opened and closed: a mobile part is pressed against a rigid outgrowth or both parts are mobile and specular, meeting on the central axis. The edges are mostly sharp with edges, bulges or knobs to prevent the grip from slipping (Nachtigall 2005).

Example	3. Odontomachus (gen. *Odontomachus*).

Odontomachus have large mandibles capable of opening 180°, locking the jaws by an internal powerful and fast mechanism used to either kill or maim the prey. Peak speeds between 126 and 230 km/h have been recorded for *Odontomachus bauri* with a peak force in the order of 300 times its own body weight. The jaws are also used as a catapult to escape a threat (Patek et al. 2006).

Load in the middle. Fulcrum and effort are at opposite ends, as in a wheelbarrow or a simple bottle opener (Sclater 2011). No examples were found in nature.

Effort in the middle. Fulcrum is at one end, load at the opposite end: as in a fishing rod (wheel is the effort, rod resting against the fisherman's trunk is the fulcrum, fish is the load) or in a pair of tweezers (Sclater 2011).

Example	1. Rove beetle (fam. *Staphylinidae*) (Dee et al. 2011)

7.2 *Inclined Plane*

A flat inclined surface can be used to do work applying a smaller force over a longer distance to lift a load vertically. The mechanical advantage equals the length of the slope to the height it spans. Although a short plane with a wide angle may do a job faster, it requires more force than a long plane with a narrow angle (Fig. 28).

Inclined plane (or wedge) is a triangle-shaped tool, which is used to lift or separate two objects (as an axe). Natural organisms broadly use inclined planes, in particular to optimize their locomotion using the surrounding medium.

Example	1. Fishes and aquatic mammals

Fishes and aquatic mammals have developed their entire body morphology on inclined planes and streamlines to avoid turbulence, which would cause great losses in kinetic energy. Streamlined bodies have a bluff shape in the front and a gradual taper at the back to prevent the flow to separate. As a result, the body is pushed forwards by the closing fluid behind it (Biewener 2007; Hertel 1963).

Example	2. Hydrofoil swimming

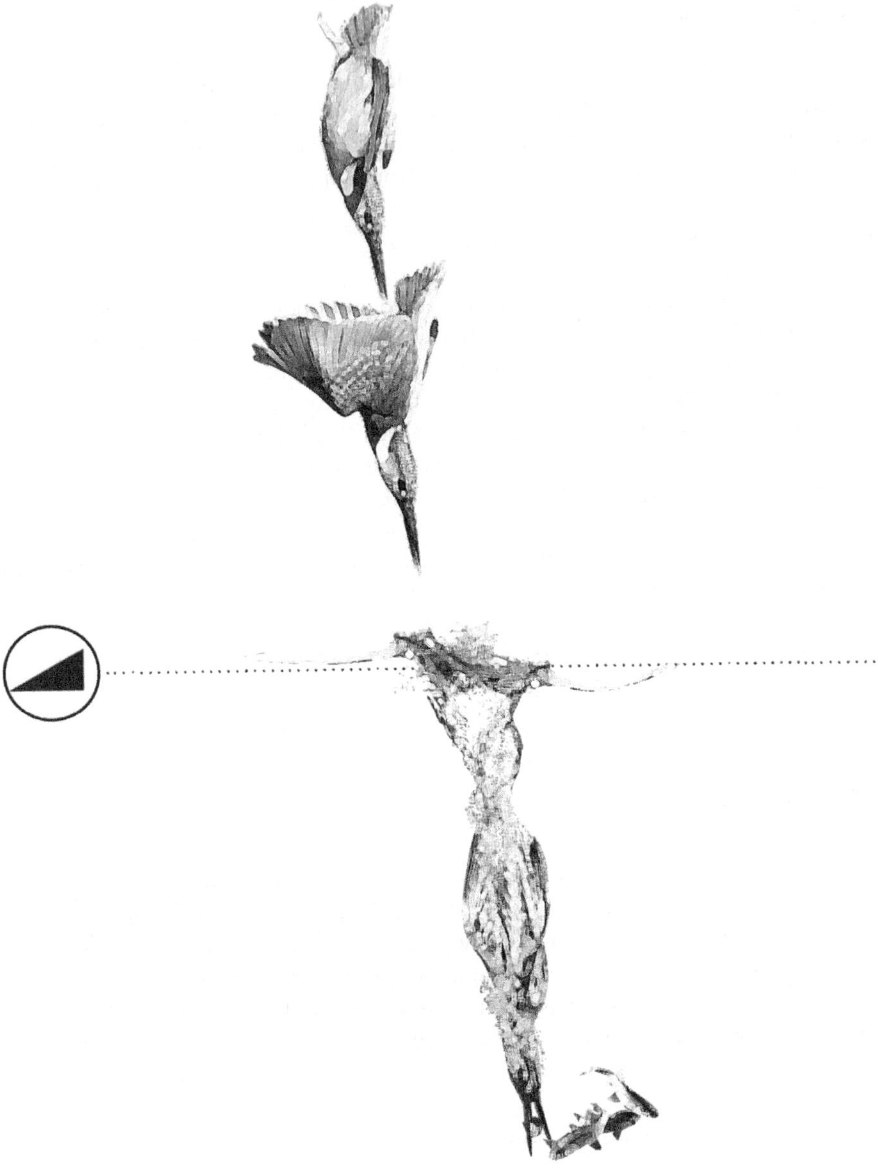

Fig. 28 The Kingfisher (sp. *Alcedo atthis*) dives down on its prey in an aerodynamic wedge shape, fending the water surface in a fraction of a second with barely a splash

Thunniform swimmers and cetaceans produce thrust by oscillating their caudal fins in a lateral or dorsoventral plane. The streamlined shape of the fins induces an asymmetric water flow over the surface resulting in a velocity differential of the flow, producing thrust (Biewener 2007).

Example	3. Sharks

Sharks have asymmetrical wedge-shaped pectoral fins and tail helping to counteract the animal's negative buoyancy by producing lift. The fins are used as 'wings' fixed in the dorsoventral plane, producing lift by separating the flow as the shark glides forward (Biewener 2007).

Example	4. Birds

To move in the air birds have adapted quite similar techniques. The wings present a curved wedge shape to avoid air viscosity and make the most of the lifting effect. As for fishes, birds' wings are shaped to prevent flow separation (and turbulence), which would stall the airfoil. In fact birds have developed specialized feathers, alulae, which bend with the airflow and help maintaining the flow attached to the wing (Biewener 2007; Hertel 1963).

Screws have a helical ridge or a narrow inclined plane, wrapped around an axe.

Example	1. Seeds

Many seeds have propeller-shaped appendages allowing them to fall in a spirally descending motion. In this case the screw plane is not used to increase the movement's efficiency but, on the contrary, to increase air friction reducing the falling speed and rising the probability of a gust of wind blowing away the seed (Fig. 29).

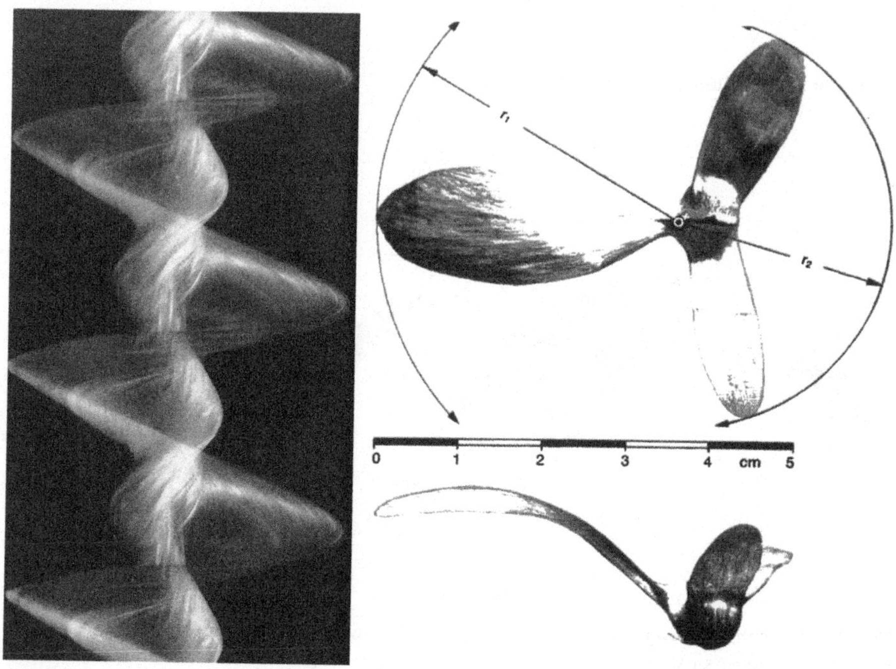

Fig. 29 Illustration and chronophotographic sequence of a seed with propeller wings falling in a screw-like motion (modified from Hertel 1963)

7.3 Joints

Joints are most important elements in moving a body made of rigid parts. They do not only facilitate movement control fixing degrees of freedom, but also permit mechanical advantage. Therefore, most rigid structures, which do not have any kind of joint (tortoise's shell), are usually used for external protection of the organism and cannot really be considered to be part of kinetic action (Fig. 30).

The design of a joint depends on its mobility and the amount of force transmitted across it. A joint with limited mobility can be simple even if the forces transmitted across it are large. In vertebrate skeletons the sliding surfaces of a joint are covered with cartilage. As cartilage has limited resistance to compression, joints exposed to high loads need larger joints to provide larger contact area (Currey 1980).

Pivot Joint—a pivot rotating within a ring, or a ring moving around a pivot.

Example	1. Humans (sp. *Homo sapiens*)

The joint between the atlas and axis (first and second cervical vertebrae), under the skull, allows turning the head from side to side. In the forearm, radius and ulna twist against the upper arm: used to unscrew the lid of a jar (Britannica 2012).

Sellar Joint—is convex in one direction and concave in the direction at right angles to the first. It transmits swing movements as flexion-extension, abduction-adduction and movements combining these: swings with rotation of the body.

Example	1. Humans (sp. *Homo sapiens*).

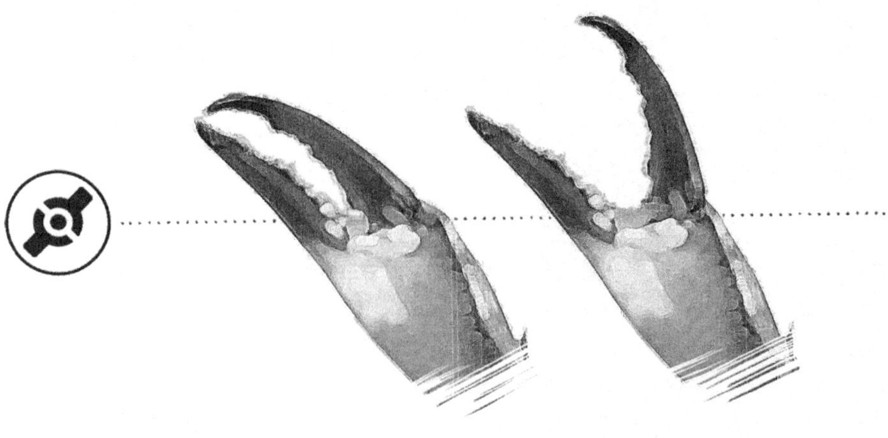

Fig. 30 A pinching movement is achieved in the crab claw through a hinge joint uniting two rigid parts of the chelae's exoskeleton

The carpometacarpal joint of the thumb can be swung from side to side or from behind to forward: especially used for opposition movements (Britannica 2012).

Ball joint—A ball joint is used for allowing free movement in two planes at the same time, including rotating. Combining two such joints with control arms enables motion in all three planes. The ball joint can be spheroidal or ellipsoidal.

The spheroidal ball joint ("ball-and-socket"), is the only joint to allow three degrees of freedom. The spherical surface of the male element moves within a depression on a female element (Nachtigall 2005).

Example	1. Wood mouse whiskers (sp. *Apodemus sylvaticus*) 2. Longhorn moth antennas (sp. *Nemophora degeerella*) 3. Harvestmen legs (or. *Opiliones*) (Nachtigall 2005)

Ellipsoidal ball has two ovoid surfaces, which are either convex in all directions or concave in all directions (like two sides of an eggshell), varies in length and in curvature. It has two movements and allows opposition only to a small degree.

Example	1. Humans (sp. *Homo sapiens*)

The joint between the second metacarpal and the first phalanx allows the finger to flex and extend, swing toward or away from its neighboring finger, and to swing forward with a slight amount of rotation (Britannica 2012).

Hinge Joint—connects two solid objects allowing a limited angle of rotation around a fixed axis. In biology many joints work as hinges, which may also be made of flexible materials or of moving components (Nachtigall 2005).

Example	1. Clam (cl. *Bivalvia*). The shell has two rigid halves, connected by a ligament and a hinge joint allowing only rotation (Britannica 2012)

Flexure Joints—are very common in nature, and very rare in technology due to the fast wear and deterioration of the materials. A flexure mechanism is designed to produce a specific motion upon application of a force, by bending of the joint.

The system relies on the bearing element being made of a material, which can be repeatedly flexed without disintegrating.

Example	1. Ladybird wings (*Coccinellidae*)

With three folding systems the wings are folded into a small package that fits under the rigid wing covers. No real hinges, as the foldings are flexible (Fig. 31).

Fig. 31 The unfolding of a ladybird's wings: the wings have no rigid joints as in bird's skeletons but the surfaces are folded through flexible hinges

7.4 Chain Systems

A system of rigid elements connected to each other by different kind of joints form a closed chain or a series of closed chains. When one link is fixed, the possible movements of the other links relative to it and to one another depend on the number of links and the number and types of joints. By varying the length of the links, the mechanism can convert uniform rotary to non-uniform rotary motion or continuous rotary to oscillatory motion (Britannica 2012; Shigley and Uicker 1995) (Fig. 32).

Roller Chain—consist of a series of short cylindrical rollers held together by side links that can move only bi-dimensionally, and are commonly used for transmitting

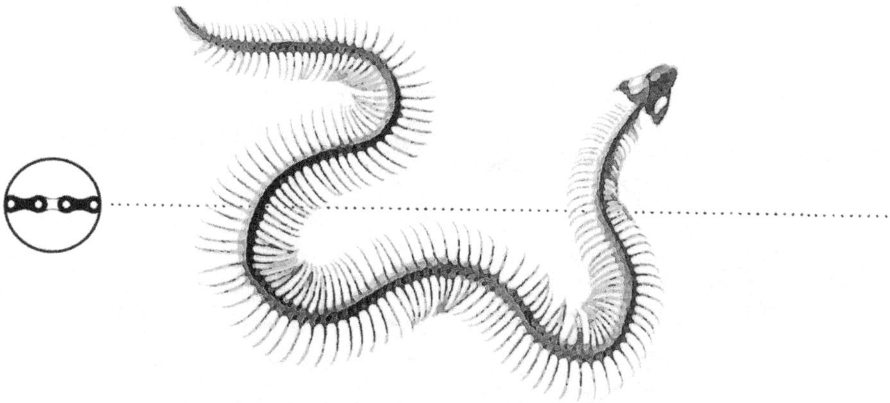

Fig. 32 A roller chain system: the spine of a snake allows deformation mainly on the animal's coronal plane, transmitting motion through the linear body by deforming into a set of sinusoidal shapes

mechanical power. In nature these systems are often supporting structures, tensed and controlled through muscles (Nachtigall 2005).

Example	1. Gecko spine (infr. or. *Gekkota*) can roll up and unroll spirally 2. Stork neck (gen. *Ciconiidae*) is extremely flexible and can bow in different S shapes

Kinematic Chain—refers to an assembly of rigid bodies connected by joints which can make very complicated movements three-dimensionally (Nachtigall 2005).

Example	1. Mullet mouthparts (fam. *Mullidae*)

8 Connection Patterns

As for the geometry patterns, the following 'family trees' of connections are represented concerning the mechanical parameters, with the systems classified according to the body structure (Figs. 33, 34, 35).

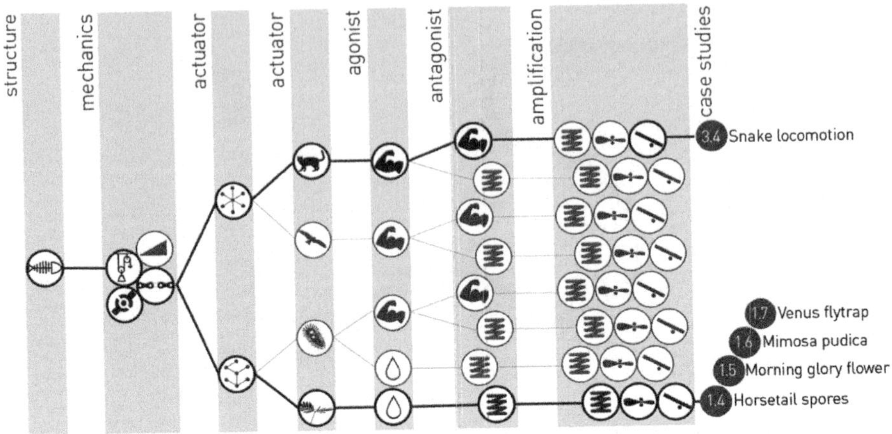

Fig. 33 Biomechanics of Motion. Connection patterns of endoskeletal body types with the other parameters studied. For case studies, refer to the deepened analysis in Chap. 6

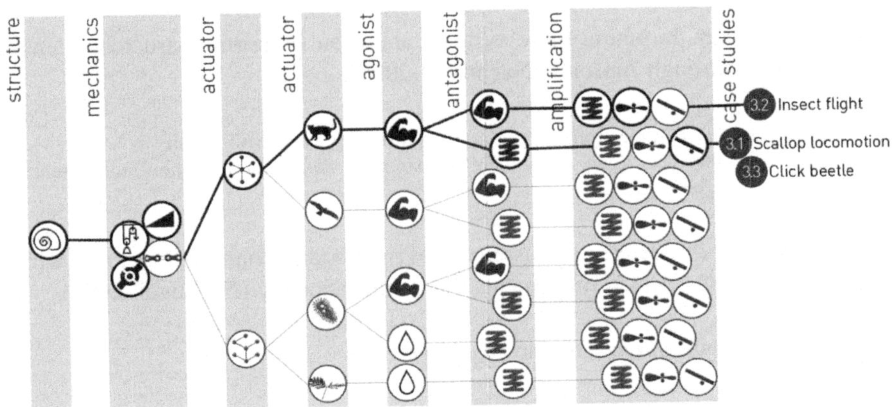

Fig. 34 Biomechanics of Motion. Connection patterns of exoskeletal body types with the other parameters studied. For case studies, refer to the deepened analysis in Chap. 6

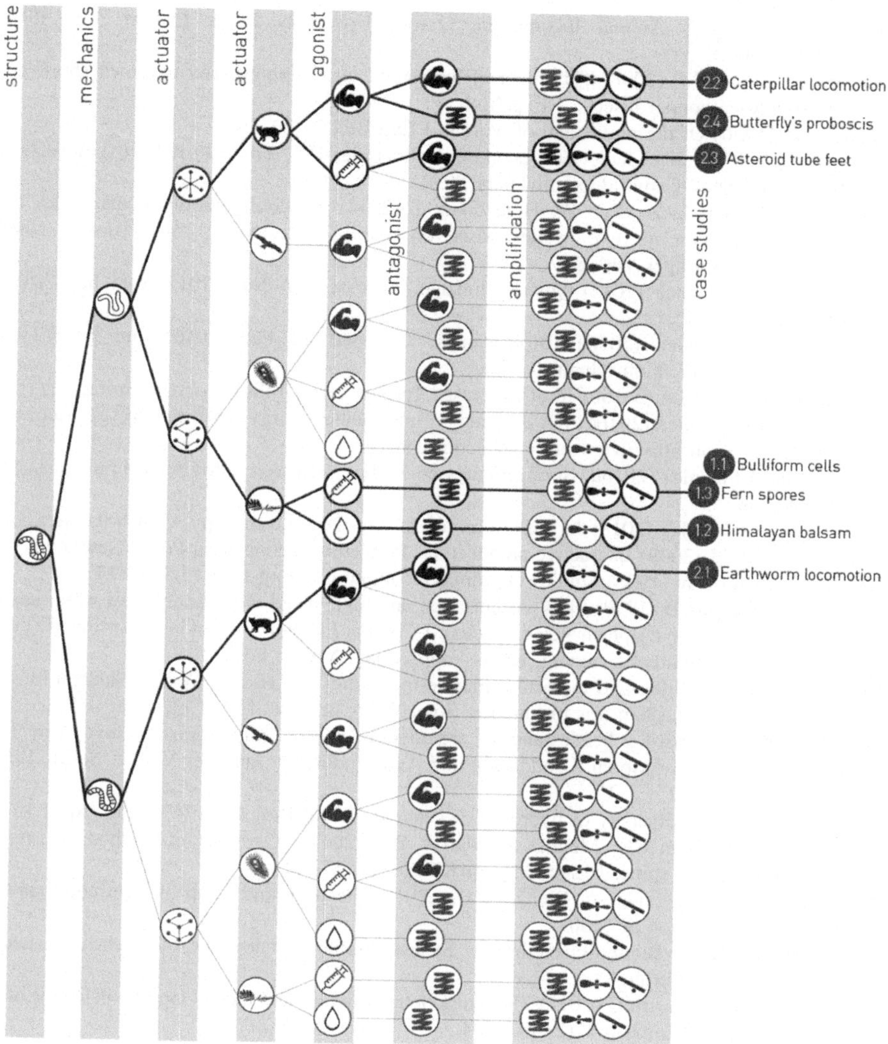

Fig. 35 Biomechanics of Motion. Connection patterns of hydrostatic body types with the other parameters studied. For case studies, refer to the deepened analysis in Chap. 6

References

Alexander RMcN (2000) Storage and release of elastic energy. In: Nigg BM, MacIntosh BR, Mester J (eds) Biomechanics and biology of movement. Human Kinetics, Champaign, Ill

Alexander RMcN (2004) Hitching a lift hydrodynamically—in swimming, flying and cycling. J Biol 3(2):7

Basili GD, Temple SA (1999) Demographic characteristics of Dickcissels in winter. Stud Avian Biol 19:281–288

Biewener AA (2007) Animal locomotion. Oxford University Press, Oxford, UK. ISBN 978-0-19-850022-3

Burrows M, Sutton GP (2008) The effect of leg length on jumping performance of short- and long-legged leafhopper insects. J Exp Biol 211:1317–1325

Chapman G (1950) Of the movement of worms. J Exp Biol 27:29–39

Currey JD (1980) Skeletal factors in locomotion. In: Elder HY, Trueman ER (eds) Aspects of animal movement. Cambridge University Press

Dee L, Kai Z, Peiping Z, Ziyu W, Hongzhang Z (2011) 3D configuration of mandibles and controlling muscles in rove beetles based on micro-CT technique. Anal Bioanal Chem 401:817–825. Springer-Verlag

Encyclopædia Britannica (2012) Muscle, linkage. Accessed 16 Nov. 2012 from http://www.britannica.com/

Encyclopadia Britannica (2016a) Biomecanics. Accessed 22 Feb. 2016 from http://www.britannica.com/topic/biomechanics-science

Encyclopadia Britannica (2016b) Joint. Accessed 16 Nov 2012 from http://www.britannica.com/

Encyclopædia Britannica (2016c) Kinetic energy, mechanical energy, potential energy. Accessed 30 Jan 2016 from http://www.britannica.com/science/

Encyclopadia Britannica (2016d) Spring. Accessed 30 Jan 2016 from http://www.britannica.com/science/spring

Encyclopadia Britannica (2018) Jerboa. Accessed 15 Jan. 2018 from http://www.britannica.com/

Fox MA, Kemp M (2009) Interactive architecture. Princeton Architectural Press, New York

Gray J, Lissmann HW (1938) Studies in animal locomotion. J Exp Biol 15:506–517

Heberholz J, Schmitz B (1999) Flow visualisation and high speed video analysis of water jets in the snapping shrimp (Alpheus heterochaelis). J Compar Physiol A 185:41–49. Springer-Verlag

Hertel H (1963) Struktur. Bewegung, Krausskopf-Verlag, Mainz, Form

Herzog W (2000) Cellular and molecular muscle mechanics. In: Herzog W (ed) Skeletal muscle mechanics: from mechanisms to function. John Wiley & Sons, Ltd., Chichester, UK

Isaak M (2003) Bombardier beetles and the argument of design, TalkOrigins Archive, May 30

Janocha H (ed) (1992) Aktoren, Grundlagen und Andwendungen. Springer-Verlag, Berlin. ISBN 978-3-662-00418-0

Kier WR (2012) The diversity of hydrostatic skeletons. J Exp Biol 215:1247–1257. April 15

Le P, Best TM, Khan SN, Mendel E, Marras WS (2017) A review of methods to assess coactivation in the spine. J Electromyogr Kinesiol 32:51–60

Marmottant P, Ponomarenko A, Bienaime´D, (2013) The walk and jump of Equisetum spores. Proc Roy Soc Biolog Sci 280

Nachtigall W (2005) Biologisches design, Systematischer Katalog fur bionisches Gestalten. Springer Verlag, Berlin

Nguyen V, Lilly B, Castro C (2014) The exoskeletal structure and tensile loading behavior of an ant neck joint. J Biomech 47(2):497–504

Nigg BM, Stefanyshyn D, Denoth J (2000) Mechanical considerations of work and energy. In: Nigg BM, MacIntosh BR, Mester J (eds) Biomechanics and biology of movement. Human kinetics. Champaign, Ill

Patek SN, Baio JE, Fisher BL, Suarez AV (2006) Multifunctionality and mechanical origins: ballistic jaw propulsion in trap-jaw ants. Proc Natl Acad Sci 103(34):12787–12792

Piper R (2007) Extraordinary animals: an encyclopedia of curious and unusual animals. Greenwood Press

Quillin KJ (1999) Kinematic scaling of locomotion by hydrostactic animals ontogeny of peristaltic crawling by the earthworm Lumbricus terrestris. J Exp Biol 202:661–674

Sclater N (2011) Mechanics and mechanical devices sourcebook (5th ed). McGraw-Hill Companies Inc., USA

Shigley JE, Uicker JJ (1995) Theory of machines and mechanisms (1st ed. 1980), 2nd ed. McGraw-Hill Inc., USA

Shigley J, Uicker J (2010) Theory of machines and mechanisms, 4th edn. Oxford University Press, USA

Stein R, Zehr EP, Bobet J (2000) Basic concepts of movement control. In: Nigg BM, MacIntosh BR, Mester J (eds) Biomechanics and biology of movement. Human Kinetics. Champaign, Ill

Thompson D'AW (2006) On growth and form. Cambridge University Press, 1961 (Trans: Albus A). Eichborn AG, Frankfurt am Mainn

Turovskiy Y, Kashtanov D, Paskhover B, Chikindas ML (2007) Quorum sensing: fact, fiction, and everything in between. Adv Appl Microbiol 62:191–234

Valentine JW (2004) On the origin of phyla. The University of Chicago Press, Chicago and London. ISBN 0-226-84548-6

Versluis M, Schmitz B, Von der Heydt A, Lohse D (2000) On the sound of snapping shrimp. Department of Physics, University of Twente, The Netherlands, Department of Zoology, Technical University Munich, German. Accessed 27 April 2016 from https://www.youtube.com/watch?v=ONQlTMUYCW4

Viegas J (2005) Kinetic and potential energy: understanding changes within physical systems. The Rosen Publishing Group, Inc., New York. ISBN 1-4042-0333-8

Vincent J (2008) Imperfection as design parameter. In: Hensel M, Menges A (eds) Morpho-ecologies. Architectural Association, London

Ward AJ, Krause J, Sumpter DJ (2012) Quorum decision-making in foraging fish shoals. Public Library of Science one (PLoS) 7(3)

Chapter 5
Rhythmics of Motion

Abstract In locomotion, the pattern and timing of the movement of limbs and core identifies different gaits, which are medium-, geometry- and outcome-specific. Different movement patterns arise to answer to changing conditions, by adapting speed, rhythm and effort during the duration of the action, modifying and balancing energy use and need of action. This chapter presents three main concepts defining variations in locomotion and five categories of locomotion, with the relative variations of gait.

Repetitive and looping motion patterns within short timeframes, typical of animal locomotion, are strongly characterized by the rhythm with which the appendages and body core take action. While aspects as timing and coordination in plant's movement are certainly not less important for the efficient adaptation of the organism, the full cycles of change occur on much longer time scales, making the need for a motion rhythm less obvious to achieve a good kinetic performance.

Rhythmics of motion deal with the aspects distinguishing specific kind of locomotion, as the medium the organism is moving in, number and type of limbs, timing and pattern of the limb interaction, etc.

Techniques and patterns of locomotion are surprisingly recurring despite the enormous diversity of systems in the animal and vegetal world. This happens because in the end, all systems need to overcome the same difficulties given by the physical principles of the medium they come into contact with. Terrestrial animals mainly deal with mass-related gravitational forces, flying animals need to generate enough lift and thrust to support their weight and move forward, aquatic animals are on one hand supported by the buoyancy effect but must overcome drag and fluid viscosity (Biewener 2007).

In order to give a comprehensive view on the argument, a few concepts regarding all kinds of locomotion, as gait, phase and coordination, are defined. Follows a classification of gaits according to body structure and medium of locomotion. As solutions in nature are often the effect of complex combinations of many aspects, radically different organisms are often found to share similar features, fitting into more than one category. With the final aim of a biomimetic technology

S. Persiani, *Biomimetics of Motion*,
https://doi.org/10.1007/978-3-319-93079-4_5

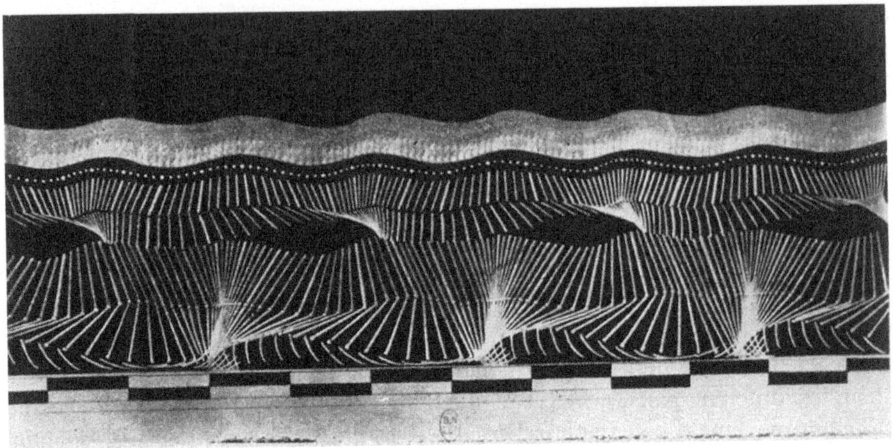

Fig. 1 Marey (1884), chronophotographic sequence of a walking man. The markings on the body identify the sequential movement of the head, the upper and lower limbs during the walking motion. The rhythmic repetition of the alternated pendulum-like movement of the limbs highlights the movement pattern typical of the biped walking gait

transfer, the following chapter aims to systematically list a broad multitude of locomotion effects rather than suggest distinct categorizations (Fig. 1).

1 Gait

Animal gaits define and characterize the timing and frequency with which the limbs support the body during the stride. As in any machine that changes gear to accelerate or decelerate, changing gait is a way for the body to adapt the same system to a progression of speeds, changing terrain, maneuvers and guarantee energetic efficiency. To reach their peak speed, animals often need to transit through several gaits (Fig. 2).

Animals move faster by increasing their stride frequency, stride length, or both. Speed changes also occur within a gait, but to move over a greater range of speeds, gait transition is necessary. The transition occurs instinctively at specific speeds, varying from individual to individual and from acceleration to deceleration. Occasional gaits are also used for uncommon and restrained movements during

Fig. 2 Simulation of the terrestrial locomotion in a salamander. The figure covers a time-span of 5.4 s all four footfall patterns are represented by the bars recording the period when the foot is on the ground (from the top: left hind, left front, right front and right hind). At approximately 1/3 of the sequence, a change in the pattern can be seen: that is a change of gait from walking to trotting (from Harischandra et al. 2011)

limited timeframes, as can be sideways and backwards maneuvering, also changing the amount of supports normally used, as in four-legged crawl for humans in tight spaces (Alexander 1981).

While locomotion is tightly related to the physical proprieties of the medium, gaits are more easily associated to the geometrical subdivision of the body in appendages, as the aim of a gait is finally to coordinate the body parts in a kinetically efficient rhythm.

1.1 Gait and Energy

One main reason for animals to change gait is that for definite speeds, specific footfall patterns are energetically more efficient. Walking at a very fast speed is strenuous for the organism, so the pattern is changed into running. The action of the muscles to accelerate or decelerate the limb is coupled with oscillating systems allowing energetic return and minimal energy input (Wilson et al. 2000) (Fig. 3).

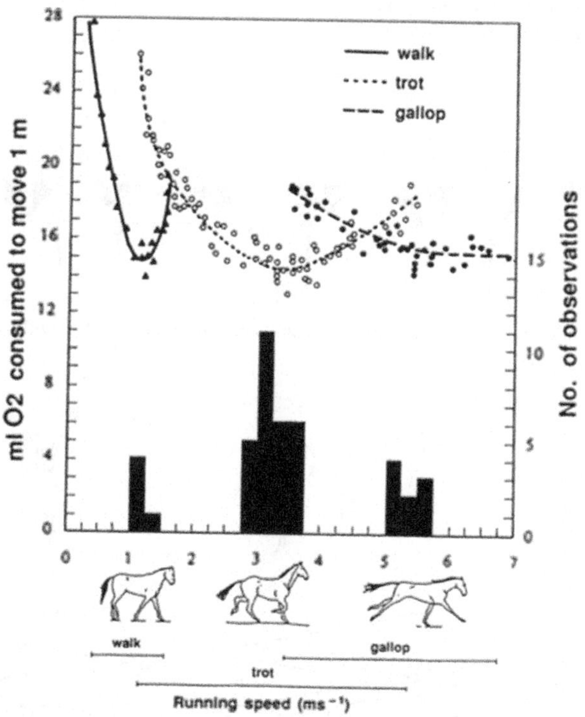

Fig. 3 Energy expenditure in three horse gaits (from Hoyt and Taylor 1981). Each gait typically has an optimal speed (at the lowest point of the curve), when the body manages maximal output at minimal effort. Above and below the optimal speed, the effort quickly rises, until the energy cost is too high, and the body switches to another gait. Gait transitions occur near speeds at which the curves intersect

Other gaits as burst swimming or in intermittent flight in birds build up energy storage and use it over a longer time. These gaits are common in smaller organisms and in media as air or water where drag is an important parameter for locomotion: bursts of energy (to overcome drag) turn out to be less energetically expensive than working continuously against it (Rayner 1981).

1.2 Gait and Equilibrium

When the bodies are in motion and possess kinetic energy, equilibrium is easier to achieve as the kinetic thrust keeps the body on its path. The more the body moves fast, the more the movement can tolerate larger departures from a state of equilibrium (Alexander 1981). This is what allows for instance bipeds to move with alternating gaits, keeping alternatively one foot off the ground. In slower running gaits, as kinetic thrust is lower than in fast running gaits, the footfall patterns are inversely symmetrical, keeping the body closer to equilibrium than in unsymmetrical and faster gaits (Minetti 2000).

Example	1. Horse (sp. *Equus caballus)* 2. Felids (fam. *Felidae)*

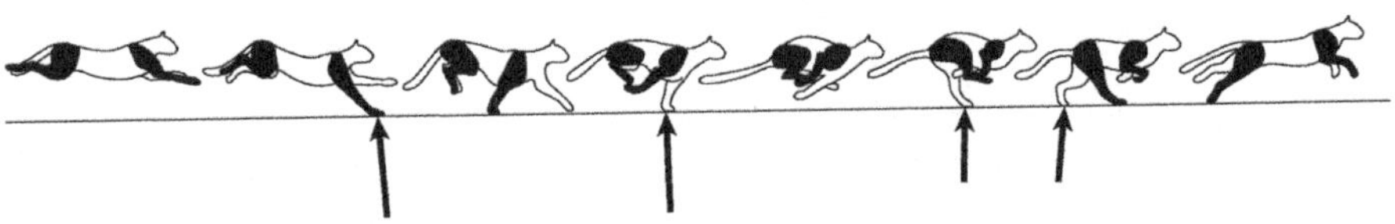

Fig. 4 Rotary gallop in a cheetah (from Bertram and Gutmann 2009). Only one or two limbs touch the ground at the same time

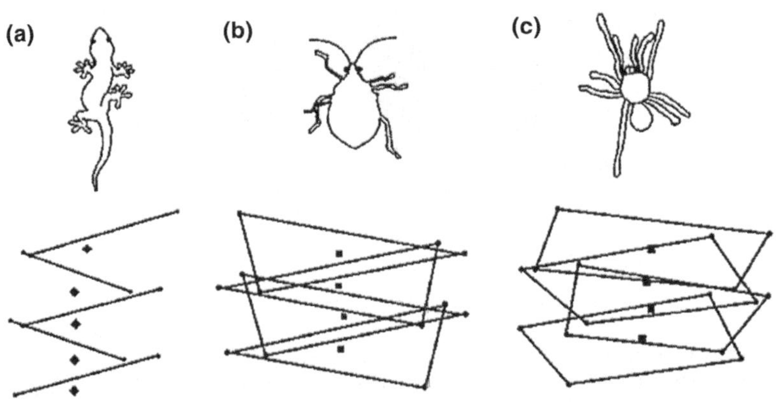

Fig. 5 Examples of slow locomotion gaits (**a**) Diagonal gait of a Gekko (sp. *Gekko gecko*); (**b**) tripod gait of a stinkbug (sp. *Erthesina fullo*); (**c**) quadrilateral gait of a spider (sp. *Ornithoctonus huwena*). The central points highlight the shifting of Centre of Mass, while the lines indicate pad-holding positions (from Ji et al. 2014)

This can be seen in trot and pace in quadrupeds, where the limbs move symmetrically in pairs of two, while in rotational gallop, which moves as a faster speeds, the limbs touch the ground in a rotational sequence, one at a time (Fig. 4).

Example	3. Ants (fam. *Formicidae*)

Ants, moving at very slow speeds use a specific gait using alternatively three of its six legs as a tripod to provide constant stability as moving forwards (see multiple limb locomotion) (Gruber 2011; Cruse and Graham 1985) (Fig. 5).

1.3 Gait and Dynamic Similarity

The speed at which the change of gait occurs depends very much on the animal's size. A child will for instance break into running at a speed at which an adult is normally still walking. Therefore, to compare animal gaits effectively, the

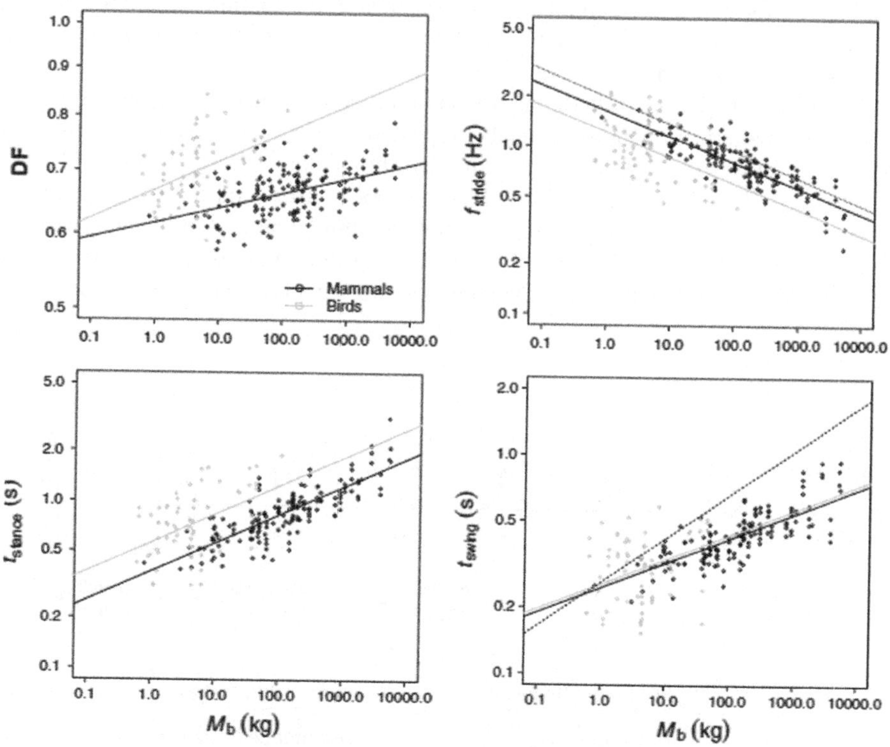

Fig. 6 Preferred gait in terrestrial vertebrates (mammals and birds) (from Lees et al. 2016). The log-log plots show four kinetic dimensions (left to right, and top to bottom: duty factor, stride frequency, stance and swing duration) against the body mass

movements must be scaled in length and time to the body-size. When scaled movements appear identical, we can say there is a *dynamic similarity*. In geometrically similar animals, dynamic similarity can often be found.

To assess similarities, the dynamic characters must be scaled in all their dimensions, which change depending on the medium and the type of locomotion. Dynamic similarity is usually assessed using the Froude number (Fr), a dimensionless number based on the speed, the limb-length and the gravitational acceleration (Biewener 2007). To calculate the Froude number however, as limb-length is an important parameter when comparing running in terrestrial animals, the Reynolds number must be taken into account where viscous forces dominate, the Strouhal numbers in the case of vibrating systems, and so on depending on the forces which dominate in a given system (Alexander 2005) (Fig. 6).

2 Phase

Any movement can be described as function of a temporal and a spatial factor. A phase is a specific and complete action occurring in a definite time lapse, which must interact very precisely and in a definite order with other phases to create specific motion patterns.

For a gait to be identified, the locomotory activity needs to enter a steady movement pattern, which is cyclically repeated in the same length of time and order. The movement of each limb is divided into a stance phase, when the foot is on the ground, and a swing phase, when the foot is lifted and moved (Fig. 7).

Any gait can be described in terms of the beginning and end of the stance phase of each limb with respect to the others. Changing the relationship between phases changes the overall locomotive movement, resulting in change in direction, in speed or in gait.

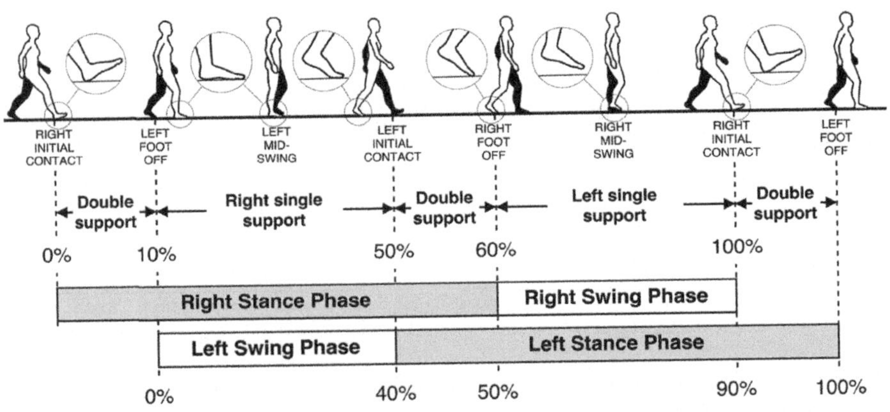

Fig. 7 Phases of a normal biped walking gait cycle (from Tunca et al. 2017)

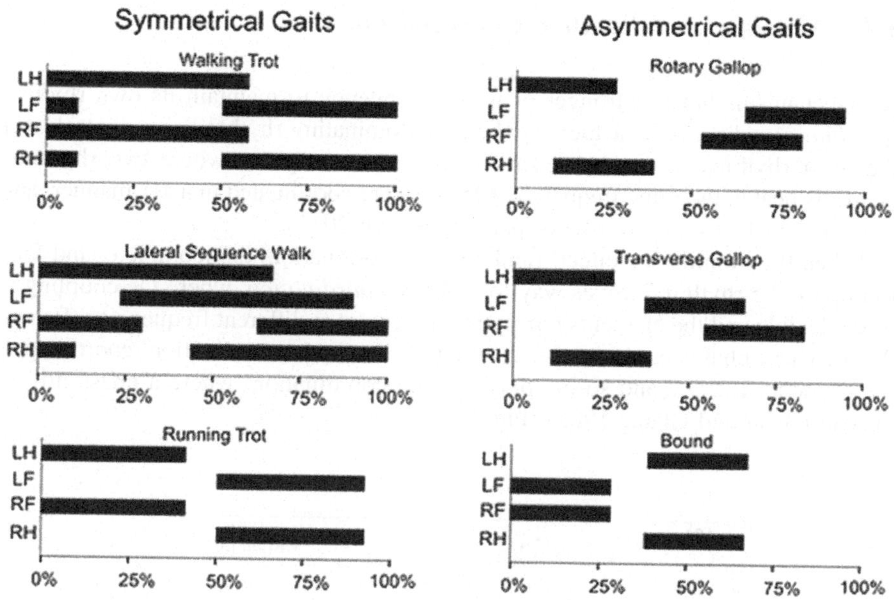

Fig. 8 Classification of six different gaits in tetrapods (from Hildebrand 1989)

In symmetrical (or alternating) gaits, as walking, trotting and swimming, the limbs in the same girdle (front limbs and hind limbs) have alternating footfall patterns accompanied by small vertical and lateral movements of the trunk.

In asymmetrical (or in-phase) gaits, as jumping and galloping, homologous limbs (both hind limbs or both forelimbs) hit the ground at the same time. These movements involve large flexion and extension movements of the spine contributing to the motion. The ipsilateral gait is a gait variation involving the synchronized motion of limbs on the same side (left limbs or right limbs), as in pacing (Miller and Schomburg 1985) (Fig. 8).

3 Coordination

Coordination is, for locomotion, a basic condition to achieve efficiency. Sudden, fast and exact movements require the body parts to responds as one, without interference or conflict. Coordination is a rhythmic behavior putting static and dynamic phases into a coherent sequence so as to obtain a harmonic movement repetition between the parts engaged in the act. The fewer parts to be activated, the simpler is the coordination task (Von Holst 1973; Altman and Kien 1985).

3.1 Absolute and Relative Coordination

All mechanisms have a stronger or weaker tendency to maintain its own rhythm. *Absolute coordination* is achieved when the dominating rhythm gains control over the other rhythms. It can be described as a constant phase where two rhythmic elements beat at the same frequency. Elements are coordinated in a 1:1 manner, and motion is locked into a repetitive pattern (Brown 2013).

When the 'dominance effect' (and perfect coordination) is incomplete, and fails to impose the rhythm, it gives way to *Relative coordination*, where the coupling is more flexible and the elements repeat the movement at different frequencies (Brown 2013). These phases can either have a kind of 'harmonic coordination' coordinating the elements 1:2, 3:2 and so on, or a 'gliding coordination' where a phase drift is present (Bush and Clarac 1985) (Fig. 9).

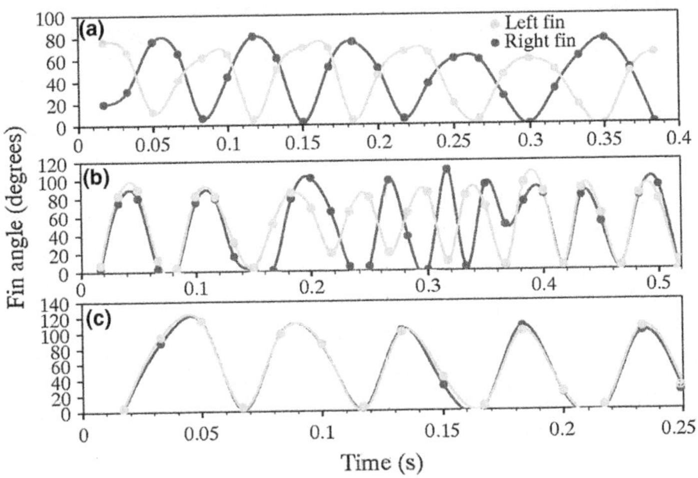

Fig. 9 Pectoral fin coordination sequences in a reef fish (sp. *Pomacentrus pavo*) (from Hale et al. 2006). (**a**) Absolute coordination with symmetrical pattern; (**b**) Relative coordination; (**c**) Absolute coordination with asymmetrical pattern

3.2 Magnet Effect

The *magnet effect*, also called entrainment, is a tendency where one rhythm (or an oscillation) acts with a sort of attraction on another rhythm influencing its frequency and drawing the phases to change timing and to couple it to the same frequency (Bush and Clarac 1985).

Example	1. Two people sitting side by side in rocking chairs (Richardson et al. 2007) will adopt, after a short time, the same rocking rhythm. The effect occurs both intentionally and unintentionally, on condition that there is a visual connection between the individuals 2. The same effect can be seen in people walking next to each other, who fall into the same walking rhythm

The use of music during coordinated motion, as treadmill walking, running and dancing, has shown to have a similar magnetic effect on the physical effort (heart rate, perseverance), providing guidance in the complex coupling of perception and movement coordination. The recurrent patterns in the acoustical structure of the music can influence the timing of the movement, in terms of synchronization of tempo and phases, and the vigor, meaning the activation and relaxation of the physical strength in a movement (Leman et al. 2013). Entrainment is widely used in physical rehabilitation, where rhythmic acoustics and music have shown to improve limb coordination of unsynchronized gait patterns (Roerdink et al. 2015; Nombela et al. 2013).

4 Passive Locomotion

Passive motion actuation exploits external kinetic factors in the surrounding medium by making the most of physical proprieties and existing turbulences in fluids as liquids and gases, and are therefore linked to the aquatic or to the aerial medium (Biewener 2007) (Fig. 10).

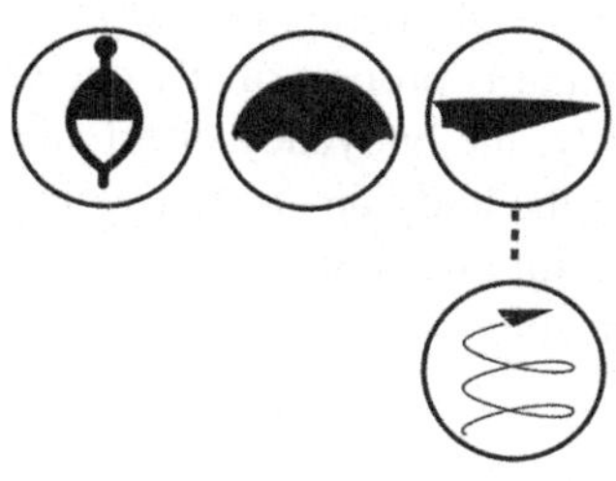

Fig. 10 Passive locomotion (from the left): floating, parachuting, and gliding—which can include the variation of soaring

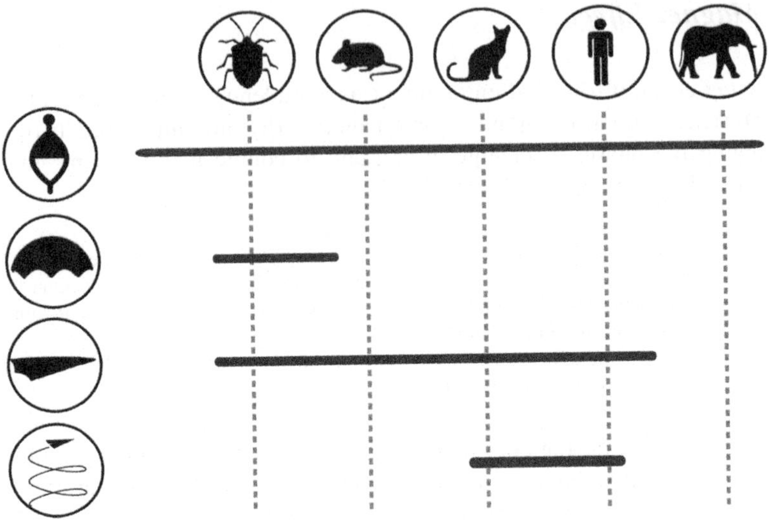

Fig. 11 Passive locomotion versus body size

In passive locomotion, organisms tend to balance the body on the edge of a stage of equilibrium between the gravitational forces and the medium's forces providing lift. The equilibrium depending on the weight of the animal and the proprieties of the medium, passive locomotion is strongly linked to body-size, and in some cases the body-size range is very narrow (Fig. 11).

4.1 Floating

Floating bodies in aquatic environments make good use of the buoyancy effect to counterbalance the gravitational forces, which are an important issue in most locomotive systems. Thanks to the upward pressure exerted by the fluid on the body, the organisms can focus their energies on forward thrust: used as a complement to many locomotion patterns, it simplifies the pattern of opposing forces by nullifying or minimizing the weight component (Fig. 12).

Since the buoyancy potential is proportionally dependent on the weight and the volume of the immersed body, the buoyancy effect can be obtained independently on the animal size, provided the body displaces a volume of liquid heavier than its own weight. Floating is therefore used in an extremely wide range of animal sizes, from micro- to macro-organisms.

Fig. 12 A mallard (sp. *Anas platyrhynchos*) floats on the water surface with no effort, under the buoyancy effect

Example	1. Marine organisms, from plankton to whales; 2. Amphibious animals; 3. Aerial and terrestrial animals adapted to aquatic environments

Because it is impossible for any macro-organism to have a density as low as that of air, floating organisms are exclusively found in aquatic environments. Microscopic organisms, as viruses, bacteria and fungi are the only ones to be light enough to float in the air.

4.2 Parachuting

Parachuting involves flying a distance mostly vertically, descending slowly. In this case, the force responsible for supporting the weight is aerodynamic drag, or air resistance. As drag is a function of the proprieties of the fluid (as density and Reynolds' numbers), size, shape and speed of the body, the smaller the body the higher are the chances to achieve sufficient drag to slow down the falling speed (Fig. 13).

In nature, parachuting mechanisms are mainly used by plants to spread seeds and spores which usually control neither movement nor direction. The distances travelled may very considerably with different weather conditions. More than 100 different kinds of parachute shapes can be identified among plants.

Propeller shapes. Some seeds have scales of various geometries which are used to increase air friction. When the fall is spirally descending, speed is reduced, raising the probability of a gust of wind blowing away the seeds.

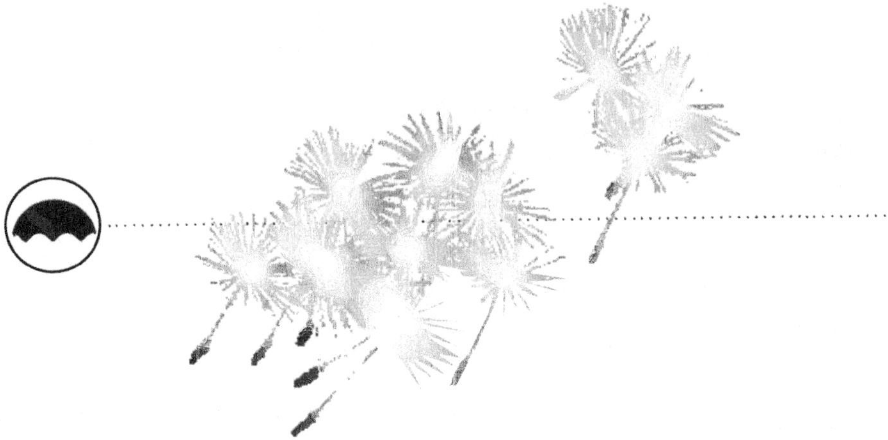

Fig. 13 Hawkbit seeds (gen. *Leontodon*) being born away by the wind. The hair-shaped umbrella maximizes air drag and friction

Example	1. Maple seeds (Gen. *Acer*)

Wad and hair-like shapes. The tuft of hairs (named pappus) on a seed, which assists the dispersal by wind may be composed of bristles or awns which offer the wind a sufficiently large surface to catch up. Radial filaments are most common.

Example	2. Hawkbit seeds (gen. *Leontodon*); 3. Valeriana (gen. *Valeriana*); 4. Willow (gen. *Salix*); 5. Epilobium (gen. *Epilobium*)

4.3 Gliding

Gliding involves 'flying' a greater distance horizontally than vertically without any use of thrust. Although the animals need to use metabolic energy to maintain the gliding position, no additional energy is used to provide thrust: gliders convert their potential energy into dynamic work so lift and drag exactly balance their weight, allowing them to descend along a fixed course at constant speed (Biewener 2007). That gliding flight is a very successful mean of locomotion is testified by the presence of gliders in every one of the vertebrate classes: fish, amphibians, reptiles, birds, and mammals (Fig. 14).

Compared to powered flight (flapping flight), gliding requires less morphological adaptation and still brings a substantial advantage to the animals that adopt it.

Fig. 14 Sugar gliders (sp. *Petaurus breviceps*) spread and tilt their limbs to extend the airfoil membrane and steer the gliding

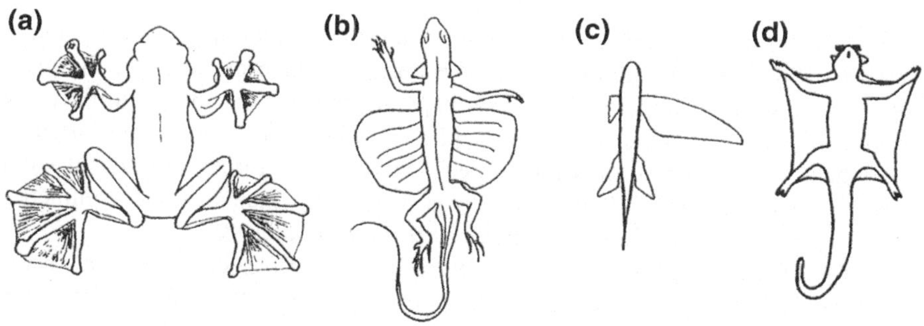

Fig. 15 Vertebrate gliders (from Day 1981)

Therefore, many animals have developed gliding as an additional technique upon their original locomotory ones (Fig. 15).

The kinetic energy necessary to overcome drag is usually obtained at the expense of potential energy with the gliding path descending relative to the air (Rayner 1981). If the body has no potential energy (as low hanging seeds), initial thrust must be provided to propel the body through the air, with a burst of motion, as when jumping. As for parachuting, also gliding flight involves some physical restrictions, limiting the size of the gliders. Three principal forces act when gliding.

- Weight (gravity—downwards);
- Lift (generated by the negative flow over the wing/fin—upwards);
- Drag (air/water-speed vector—backwards).

Further parameters to assess gliding flight are:

- Glide angle (P). The smaller P, the longer the gliding distance.
- Glide ratio (L/D). Gliding capacity is gliding distance (L) to height loss (D).

- Aspect ratio (AR) is a key parameter, which affects the lift-to-drag performance of an airfoil and is given by the ratio of the wingspan (tip to tip) to their average width. Long narrow wings with high AR values enhance lift; short stubby wings with low AR values cannot glide but are more maneuverable.

For aerodynamic reasons only large and medium-sized birds glide. Bigger birds characterized by high AR values (L/D 10–18) glide. Medium-sized birds can glide for shorter distances while smaller birds, more agile but with low AR value (L/D 2–8) cannot glide at all and must fly fast to produce enough lift. Similar happens for insects (insects with L/D < 2 are not good gliders).

Example	1. Wandering albatross (sp. *Diomeda exulans*): P3°; L/D 20; AR 15 2. Hawks and vultures (fam. *Accipitridae*): P4-6°; L/D 10–15; AR 7.5 3. Sparrow (fam. *Passeridae*): AR: 5.5 4. Dragonfly (fam. *Libellulidae*); AR: 5.1 5. Swans (gen. *Cygnus*) and geese (sub. fam. *Anserini*) are large birds, which do not glide as probably the relative payload is too heavy

High wing loadings give faster gliding motions with larger turning radius (Norberg 1981). Normal gliding of birds requires less than one third of the metabolic rate as during flapping flight. This happens because on one hand, only few muscles are involved in the control of the gliding motion, on the other muscles need to produce only isometric or semi-isometric contraction (force is produced while muscle remains the same length) which is less costly than isotonic contraction (force is generated and muscle is lengthened) (Goldspink 1981).

As for animals gliding in air, sharks have developed particular pectoral fins to use dynamic lift. The major drawback of this system is that these fishes must continuously move to stay afloat, while it also makes it impossible for them to swim backwards.

Example	6. Sharks (Sup.or. *Selachimorpha*)

4.4 Soaring

Soaring is a variation of the gliding flight, which involves the additional use of the available energy in natural air-movement patterns as temperature (thermal soaring), pressure (slope soaring) or air speed (dynamic soaring) to fly upwards without any thrust. While all soaring animals can glide, very few animals that have evolved gliding can soar. This is because while gliding can be an occasional and temporary method of locomotion, soaring is used for long-distance and long-lasting flight, and is therefore predominantly used by big birds. Soaring requires large wing areas with low wing loading to give a slow gliding motion with low sinking speed (Norberg

Fig. 16 A Gannet (gen. *Morus*) soaring

1981). The soaring techniques are strongly connected with topography and consequent weather conditions (Fig. 16).

Thermal soaring uses columns of rising warm air formed on the ground through the warming of the surface by sunlight. Once a thermal is encountered, the bird flies in circles to keep within the thermal, descending with respect to the air but gaining altitude with respect to the ground before flying off to the next thermal.

Example	1. Andean condor (sp. *Vultur gryphus*)

Slope soaring (or ridge lift), is caused by rising air on the windward side of a slope, a cliff or an ocean wave, eventually augmented by thermals. With a steady wind blow, birds using the lift can virtually stay for an unlimited time aloft.

Example	2. Seabirds

In *dynamic soaring*, energy is gained by repeatedly crossing the boundary between air masses of different horizontal velocity rather than by rising air.

Example	3. Wandering albatross (sp. *Diomeda exulans*)

5 Limbless Locomotion

Slender organisms such as sperms, eels, snakes propel themselves by propagating flexure waves along the length of their body. The in-plane alternated lateral forces generated when the animal braces the sides in a rhythmic pattern of muscular contraction drives the body forward. The sequence actuation patterns are usually linked to body-type and body size (Biewener 2007; Dickinson et al. 2000) (Figs. 17 and 18).

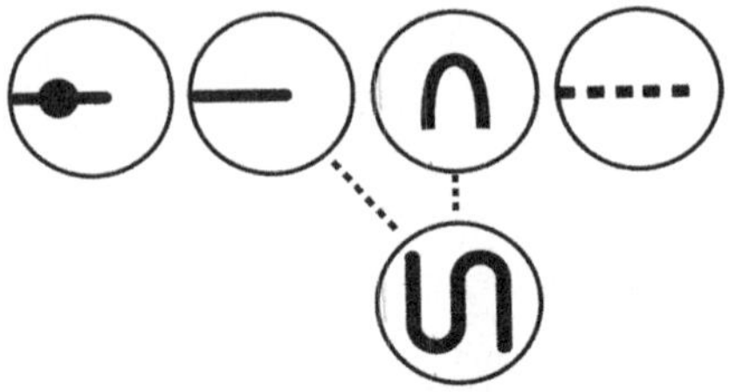

Fig. 17 Limbless locomotion (top left): peristaltic; rectilinear and concertina can be used as different gaits together with undulatory (bottom); jet propulsion

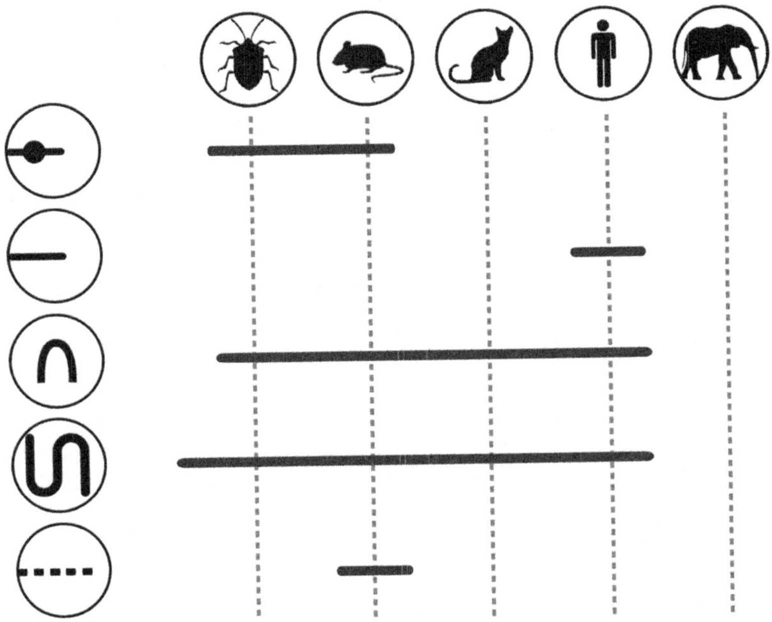

Fig. 18 Limbless locomotion versus body size

5.1 Peristaltic Locomotion

Peristaltic locomotion involves a series of wavelike contractions and relaxations of the muscles around the circumference of the body. In locomotion, the contraction wave occurs in the opposite direction of the thrust (Valentine 2004; Seymour 1969) while in the visceral functions, as in the esophagus, the contraction occurs in the same direction as the motion, pushing food through the tract. Although the muscles contract in rapid sequence, the resulting movement is slow and undramatic. The relationship between the parietal muscle force, the volume of the tube's section and the weight of the contents limits the size of the organs performing peristaltic movements (Figs. 19 and 20).

Example	1. Human gastrointestinal tract; 2. Earthworm (sp. *Lumbricus terrestris*)

The Earthworm's peristaltic locomotion has been the object of many studies (see also case study in Chap. 6). Its hydrostatic skeleton moves by radially constricting the anterior portion of its body, creating hydrostatic pressure resulting in an increase in length. This constricted region propagates posteriorly along the worm's body. As a result, each segment is extended forward, then relaxes and re-contacts the substrate.

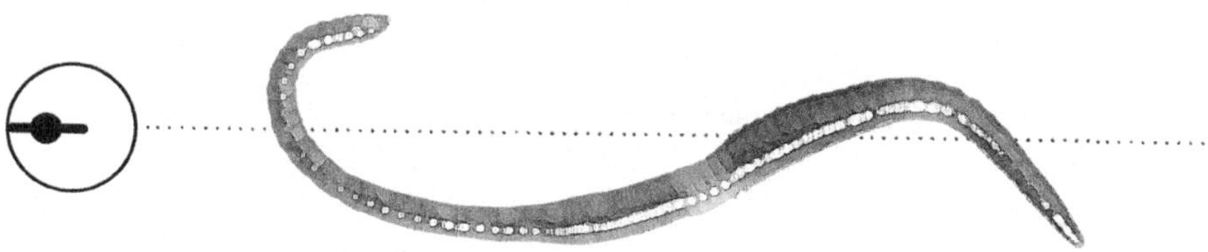

Fig. 19 Peristaltic locomotion is typical of earthworms

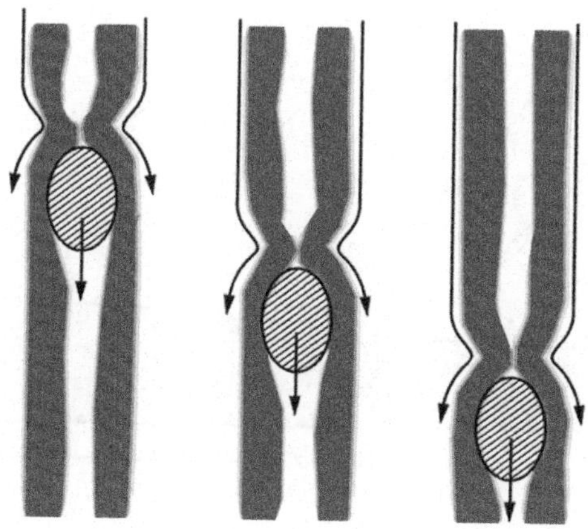

Fig. 20 Preistalsis moving the food through the digestive tract with alternating waves of muscle contraction and relaxation

5.2 Rectilinear Locomotion

Rectilinear locomotion usually refers to a gait used by snakes, where instead of bending the body, the animal progresses linearly by contracting the costcutaneous inferior and superior muscles opposing each other in waves passing in the posterior direction. Analogies with the peristaltic locomotion of the earthworm can be seen in the contraction patterns and the mechanical forces used during locomotion (Lissmann 1949). Local body lifting increases speed, reducing friction dissipation and benefitting from vertical motion as in tetrapod walking (Marvi et al. 2013) (Figs. 21 and 22).

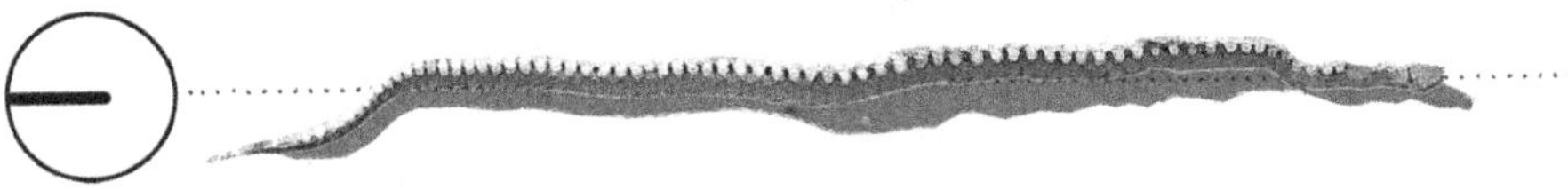

Fig. 21 Top view of a snake performing rectilinear locomotion

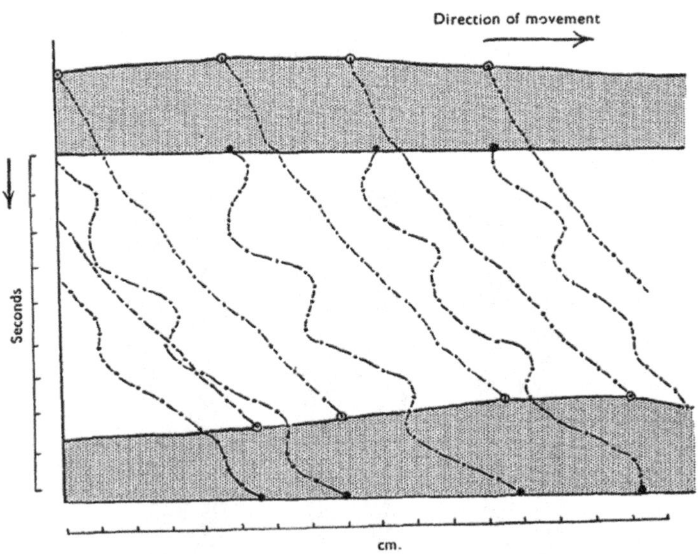

Fig. 22 Sectional scheme of a *Boa constrictor* during rectilinear locomotion (from Lissmann 1949). All parts rigidly connected to the vertebral column move forward at a relatively constant speed, while the ventral surface moves forward in a step-like motion

Rectilinear locomotion is slow and silent, and is mostly used by big snakes to stalk prey (Gans 1986). It is also used when the terrain is narrow and does not allow for sinusoidal movements. Being a form of vertebrate locomotion, it can be considered to be quite unique, as it does not make any use of the skeletal structure as a lever to move forwards (Lissmann 1949).

Example	1. Boa (sp. *Boa occidentalis*); 2. Python (gen. *Python*)

5.3 Concertina

Concertina is a gait used by many limbless species that cannot effectively undulate. The front part of the body fixes with the surface while the rear part curves to fix next to it. With the posterior providing support, the anterior portion of the body is extended to find a new grip, shifting the center of gravity forward. The movement is very simple requiring the control of one part of the body at a time and with fewer curves than lateral undulation (Biewener 2007) (Fig. 23).

Example	1. Leech (sub. cl. *Hirudinea*) 2. Geometidae caterpillar (fam. *Geometridae*)

Concertina is thought to be a transitional locomotion state in the shift from quadrupedal to limbless. Even some quadrupeds can occasionally use concertina (Fig. 24).

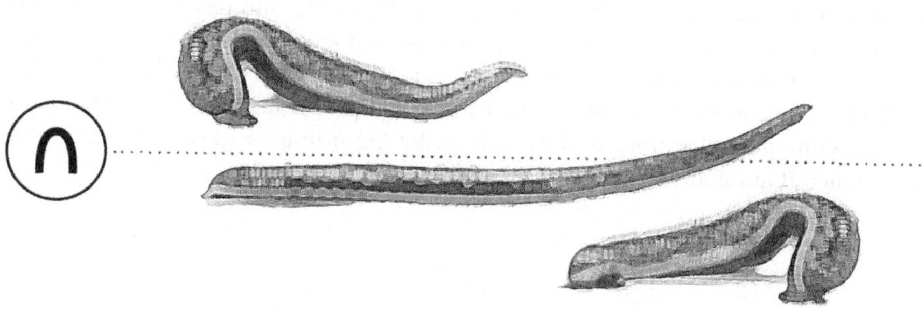

Fig. 23 Concertina in a leech walking

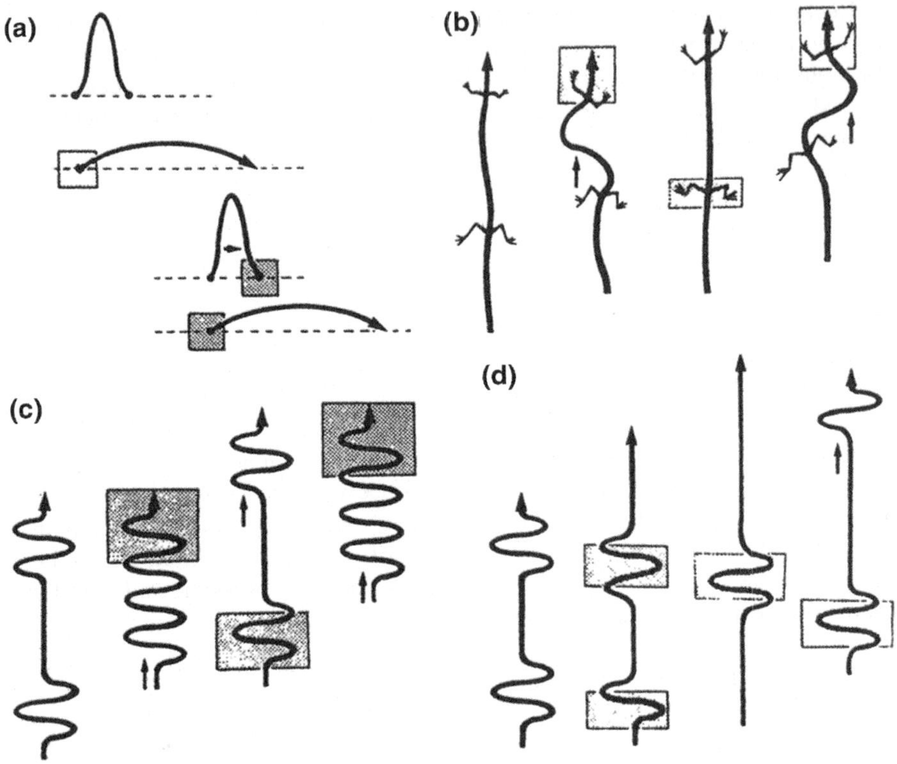

Fig. 24 Concertina in (**a**) a leech, (**b**) a lizard, (**c**) and (**d**) a snake (from Bush and Clarac 1985)

5.4 Undulatory Locomotion

Coordinated undulations of the body are produced by alternating left and right contraction waves. Undulation is a common phenomenon and has been retained as a shared primitive characteristic in many tetrapods, as lizards and crocodiles which pull their limbs against the side of their trunk and propel themselves by undulation when swimming. Therefore it is thought to be the primitive movement scheme in vertebrates (Guo and Mahadevan 2008; Biewener 2007) (Fig. 25).

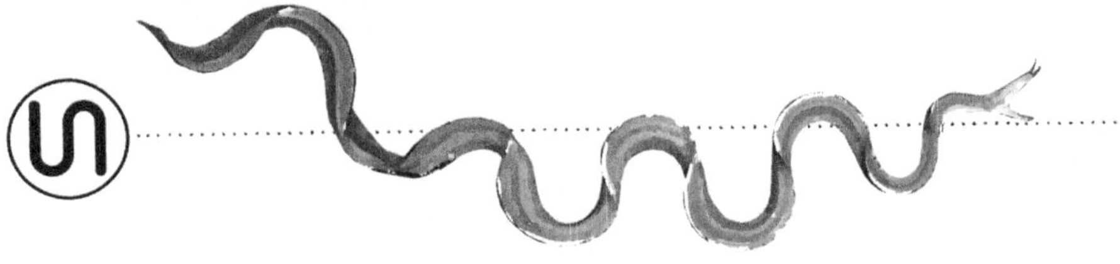

Fig. 25 Undulatory swimming in a ribbon eel (sp. *Rhinomuraena quaesita*)

The notochord has a major role in bending and the arrangement of the vertebrae provides resistance to compressive loads. The segmentation of the column influences the undulation's phase lag setting the wave length of the undulation (Valentine 2004; Gans 1985; Robert 1981; Soffe 1985).

Lateral undulation (also called serpentine locomotion or slithering) involves that the body is thrown into sinusoidal lateral waves travelling down the body axis. Whenever a bend contacts a rough or protruding surface, it locally deforms around it to provide grip. The lateral force produced on the medium (kinetic friction) is compensated by symmetrical oscillations as the body bends in opposite directions creating a resulting forward thrust. The long-limbed body follows the path established by the head, as in a roller chain. Lateral undulation is common in elongated body forms.

Example	1. Snake (sub. or. *Serpentes*) 2. In anguilliform swimmers, the amplitude of the body wave increases as it passes to the tail; 3. Crocodiles (sub. Fam. *Crocdylinae*)

Undulatory variations typical of snake gaits include slide-pushing and side-winding. In *slide-pushing*, irregular bends of the body produce vertical force on the ground. It is used for sudden movements, as when the snake is startled. Although the body might slip on the surface, the force exerted is vigorous enough for it to move. *Sidewinding* is used on slippery surfaces and differs from the lateral undulation in that the body travels diagonally as segments of the body are lifted off the ground, while the bends step on the ground exerting static friction (Moon 2001).

Variations of undulatory swimming locomotion include *Sub-carangiform*, *Carrangiform* and *Thunniform* swimming, where *only* the posterior part of the body undulates (respectively posterior half, third and fifth) propelling the fish forward. Velocity is increased by changing frequency and amplitude of the movements (Robert 1981). These movements are mostly used by active fast fishes with fusiform bodies that minimize water displacement and turbulence effects (Zug 2017) (Fig. 26).

Example	1. Salmons (fam. *Salmonidae*): Sub-carangiform swimming; 2. Cod (Gen. *Gadus*): Carrangiform swimming; 3. Tunas (tr. *Thunnini*): Thunniform swimming;

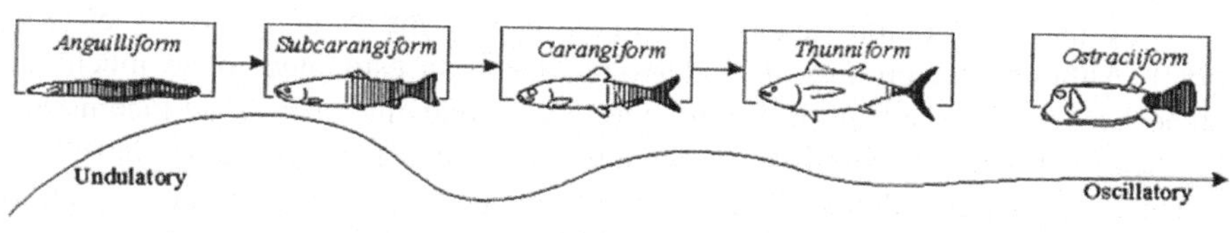

Fig. 26 Body/caudal fin propulsion in swimming (from Lindsay 1978)

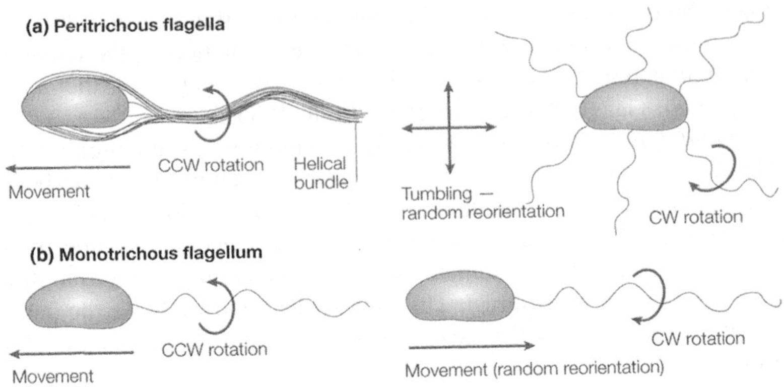

Fig. 27 Motion patterns in two different types of flagella. Counter-clockwise (CCW) flagellar rotation results in unidirectional forward corkscrew-like propulsion and a smooth-swimming motion (with the formation of a helical bundle in Peritrichous flagella). With clockwise (CW) rotation, unbundling and a tumbling motion occurs when there is a helical bundle, and a backwards propulsion and random reorientation is achieved in monotrichous flagella (from Butler and Camilli 2005)

Flagellar swimming is achieved through helical or propagated bending waves. It is used by microorganisms, which need to overcome viscous forces while inertia can be ignored (Biewener 2007). Changes in the swimming patterns, through change in waveform, amplitude and direction of motion, are considered necessary for an efficient fertilization. These changes are triggered by the presence or absence of specific chemical stimulus, which can result in swimming in a fairly steady direction, a change in direction or even a tumbling chaotic motion due to the reversal of the flagellar rotation every few seconds (Shiba et al. 2014) (Fig. 27).

Example	1. Eukaryote flagella bend the flagellum 2. Prokaryotic flagella rotate their base passing waves along their length

5.5 *Jet Propulsion*

A muscle lined wall encloses a fluid-filled chamber. Thrust is achieved in reaction to the momentum of the fluid discharged by the jet. Most of the propulsion effect is given by the acceleration of the fluid mass, not by the velocity of the jet (Kier 2012). Although jet propulsion underwater has been estimated to be much less efficient than tail propulsion in fish, due to low Froude efficiency of jet locomotion, the change of medium can visibly make a big difference as jetting squids have been estimated to achieve almost 50 BL/s (body length per second), while swimming speeds in fish have been recorded up to 25 BL/s (Fig. 28).

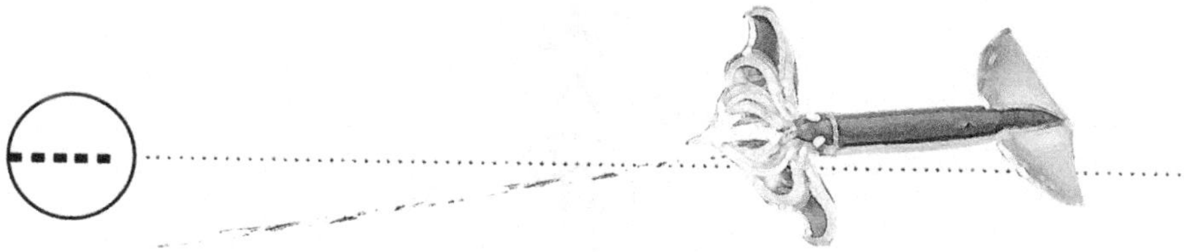

Fig. 28 Jet propulsion in a young osceanic squid (fam. *Ommastrephidae*)

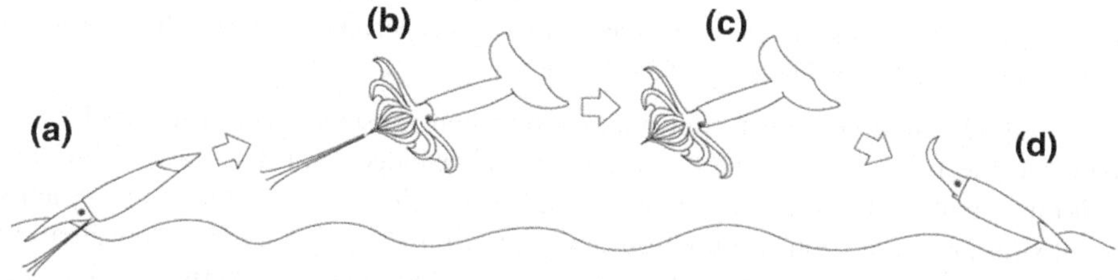

Fig. 29 Four phases in the airborne jetting of a young osceanic squid (fam. *Ommastrephidae*). Under the water surface the squid draws in seawater, (**a**) folds its arms and fins and energetically compresses the mantle producing jet propulsion and breaking the water surface; (**b**) spreads its fins and arms provided with an interstitial gliding membrane and keeps jetting until it is exhausted; (**c**) keeps gliding as long as possible though the air before (**d**) diving back into the water, folding again the arms and the fins to minimize the impact with the water surface and take advantage of the kinetic thrust it still possesses (from Muramatsu et al. 2013)

Example	1. Nautilus (fam. Nautilidae) swim by drawing water in and out of the living chamber; 2. Squids (Or. *Teuthida*) can reach speeds up to 8 m/s

The jet propulsion can clearly be seen in the case of flying squid, where the locomotory cycle can be described through four main phases: launching, jetting, gliding and diving (Fig. 29).

6 Tetrapod Locomotion

Tetrapods are four-legged vertebrates which have invaded land developing limbs and new coordination schemes upon the original undulatory one: bipeds are tetrapods who have evolved walking mainly on their back limbs, birds have evolved flight and snake-like vertebrates are thought to have lost their appendages through evolution (refer to the limbless category) (Biewener 2007).

To move at different speeds and with varying accuracy, tetrapods have evolved different footfall patterns, each identifying a different gait: as walking, trotting, canter, gallop (Fig. 30).

Fig. 30 Tetrapod locomotory gaits (top to bottom: walking, hopping, running, gallop), of which animals can use two or more, generally to change speed

Tetrapody involves great coordination between muscles (within the limbs and between the limbs) as well as coordination and balance with the trunk. While the column is stiffened to transmit forces, complex muscle systems achieve movements as flexion and extension, protraction and retraction (Massion 1985). Coordination of tail movements can participate in the overall locomotive pattern, as swinging movements of the tail that can contribute with a balancing momentum (Minetti 2000; Gans 1985; Miller and Schomburg 1985; Day 1981).

Similarities between different tetrapod species are recurring, although each type of gait is adapted to the specific body-features. Animals of different sizes for instance use take longer strides for higher speeds but in proportion similar stride lengths. Also, in mammals, change in gait tends to occur at equal Froude numbers: these mostly change from trot to canter around 2 Fr and from canter to gallop around 4 Fr (Alexander 1981). Also, different animals will use different gaits at comparable speeds—man walks while kangaroos hop—although the same gaits can occasionally be used by both species: men can hop, and kangaroos can walk on all four with the help of their tail (Fig. 31).

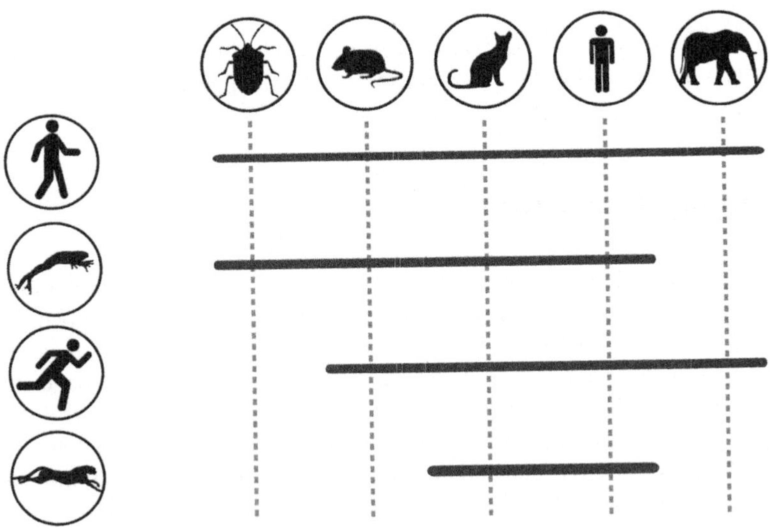

Fig. 31 Quadruped gaits versus animal size. Gaits are broadly shared between quadrupeds, with few exceptions, where body geometry and proportions do not allow specific gaits

6.1 Basic—Walking and Brachiation

When walking, animals use an inverted pendulum mechanism exchanging Center of Mass (CoM) potential energy with kinetic energy (KE), saving energy and reducing muscle work (Biewener 2006, Wilson et al. 2000). The movement is based on a pivoting system with a limb as spacer element, a joint for pivoting (ankle or wrist) and a plantar (foot or hand). As each half-swing incorporates energy for the second half swing up to 70% of the energy can be saved. This moving pattern can only be achieved at slow speeds (Fig. 32).

Example	1. Brachiation: Arboreal monkeys 2. Biped walking: Humans (sp. *Homo sapiens*) 0–9 km/h 3. Biped walking: Ostrich (sp. *Struthio camelus*) 0–8 km/h

Biped walking involves overlapping periods of support because only two limbs are involved and static equilibrium cannot be achieved. Humans change gait around 2.0 m/s from walking to running, though 3.0 m/s is considered the limit speed for walking: the gait change occurs at a slower rate since it is energetically more convenient for the organism to run over 2.0 m/s (Minetti 2000) (Fig. 33).

In a broad range of legged animals, the transition from an inverted pendulum-walking gait to a running gait occurs at dynamically similar speed, around 0.5 Fr (Rubenson et al. 2004).

Quadrupedal walking involves periods during which three limbs are always in contact with the ground to provide static equilibrium.

Example	1. Cat (sp. *Felis silvestris*)

Pentapedal walking is a rare variation of quadruped walking recalling aspects of tripod and concertina locomotion, and is mainly used by kangaroos. Having a front–hind limb asymmetry geometrically constraining the walk by the unequal length of their limbs, these animals use their big tail as a fifth limb. The strong tail, having probably evolved to counterbalance the body during hopping, is not wasted during the

Fig. 32 Three sequences of a walking man

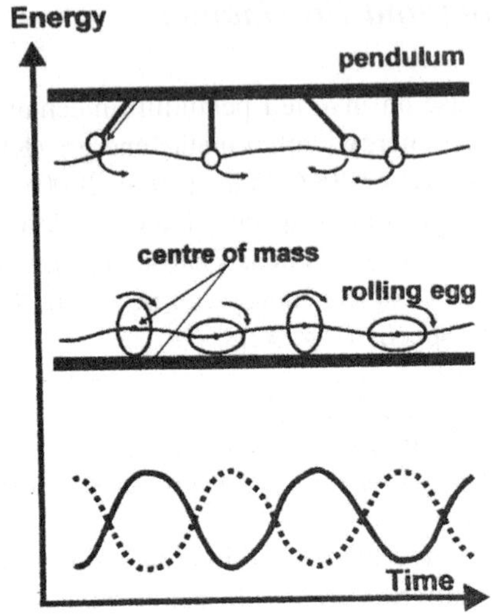

Fig. 33 Representation of the typical pendulum movement of biped walking (from Minetti 2000)

Fig. 34 Phases of the pentapedal locomotion in a red kangaroo. With its mass distributed towards the hindlimbs, the kangaroos use the tail in combination with the forelimbs, to counterbalance the unevenly distributed weight by creating a tripod stabilizing the body and providing thrust with the tail while both backlimbs are swinged forwards (from O'Connor et al. 2014)

walking motion as it provides stability and a substantial part of the mechanical work, while front and hind legs are moved in pairs (O'Connor et al. 2014) (Fig. 34).

Example	1. Red kangaroo (sp. *Macropus rufus*).

6.2 Sustained—Running, Trotting, Pacing

When running, a spring-like effect is added to the spacer-pivot-plantar system as kinetic energy is absorbed and retransmitted as potential energy through the elastic proprieties of the trunk, linkages, tendons and passive muscle elasticity (Biewener 2007; Alexander 2000) (Fig. 35).

In running, the center of gravity vaults over the legs creating a concave pendulum bouncing motion (Wilson et al. 2000) and can typically be distinguished from slower walking gaits through the presence of a flight phase during which none of the limbs touches the ground, propelling the body though the air (Fig. 36).

Fig. 35 One sequence of a running man

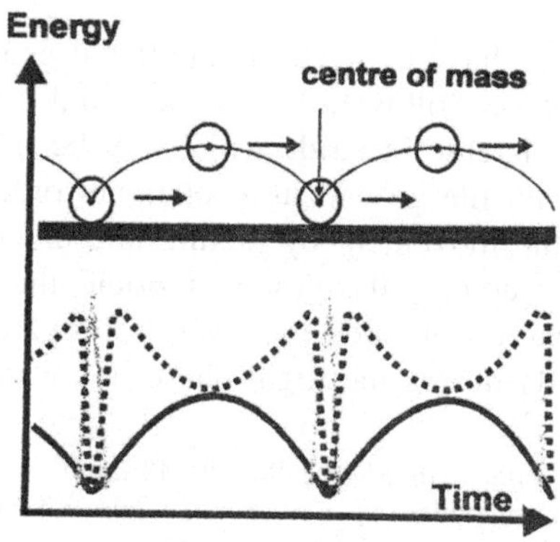

Fig. 36 Representation of the typical bouncing movement in biped running (from Minetti 2000)

Running is mostly used as a fast gait by bipeds. It involves that both feet are off the ground at regular points during one cycle, it requires more energy per unit distance than walking and is therefore less efficient in terms of cal/m, though it is faster. However, also some tetrapods become occasional bipeds when running at high speeds (Alexander 2004).

Example	1. Basilisks (Gen. *Basiliscus*) run for short distances on water; 2. Ostrich (sp. *Struthio camelus*) run up to 70 km/h

Ostriches change gait from an inverted pendulum-like walking gait (see previous part on walking) to a bouncing grounded running gait, before adopting an aerial running gait at around 3 m/s. Much of the energy involved in the transition to the aerial phase is determined by the spring energy provided by the limb (Rubenson et al. 2004).

Trotting is used by quadrupeds: diagonal fore- and hind-limb are moved in unison and each sequence is alternated with the contralateral limbs, the support periods between the strides overlap.

Example	3. Horse (sp. *Equus caballus*) trot around 13 km/h

Pacing is a (rare) trotting quadruped gait involving use of ipsilateral fore- and hind-limbs in the same phase. Due to its symmetrical component, the movement is quite unstable and results in a rocking motion (Biewener 2007).

Example	4. Camels (Gen. *Camelus*) naturally pace. Horses can be trained to, but the gait is for them unnatural

6.3 Fast—Galloping

Many four-legged animals have evolved an additional gear, gallop. The two fore limbs move almost in phase, followed by the two hind limbs. Increased speed is mainly achieved by an increase in stride which can be further increased through rotational gallop (limbs hit the ground in a rotational order), as the whole spinal column participates in the movement by flexing and extending. Short sequenced periods where the limbs contact the ground reorient the vector of the animal's trajectory (from downwards forwards to upwards forwards), as the mass of the animal travels ballistically during the flight phase (Hudson et al. 2012) (Fig. 37).

Example	1. Horse (sp. *Equus caballus)* gallop 40–48 km/h 2. Freshwater crocodiles (Sp. *Crocodylus johnstoni*) gallop 7–17 km/h, the speed depending on the size of the animal. This gait is thought to allow fast motion over irregular surfaces (Webb and Gans 1982)

Among the main aspects that are considered to influence galloping speed:

1. Swing time;
2. Stride length;
3. Peak force the limbs can withstand;
4. Muscular power available for external work.

Example	3. Cheetah (sp. *Acinonyx jubatus*) speed up to 110 km/h

Comparing the footfall patterns of two fast galloping animals, a greyhound and a cheetah, it is visible how the cheetah achieves longer strides and the shorter swing times, contributing to achieve faster speeds than the greyhound (Hudson et al. 2012) (Fig. 38).

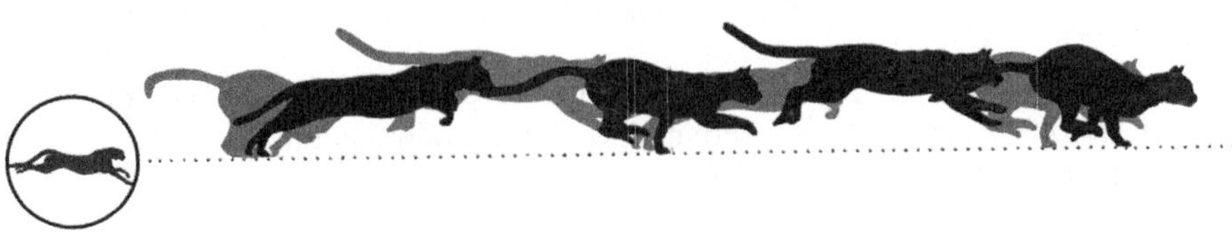

Fig. 37 Two sequences of a galloping cat

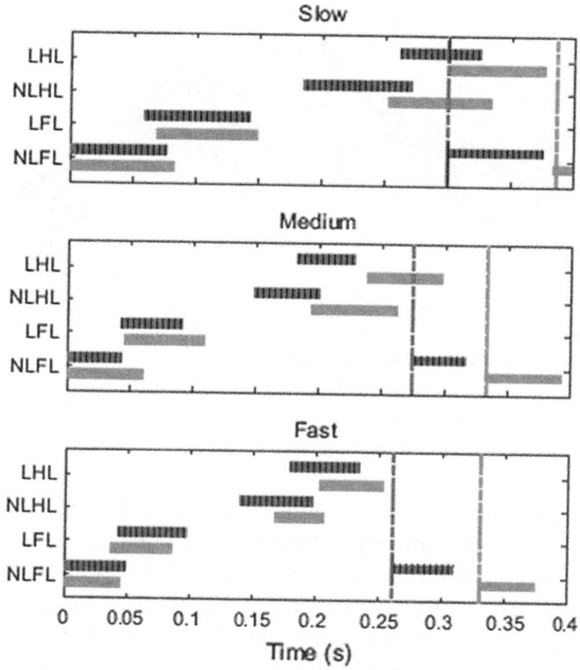

Fig. 38 Comparison of the stance timing during rotary galloping strides in a greyhound (dark grey) and a cheetah (light grey) at three different speeds. The dashed vertical lines represent the end of the stride. The order in which the limbs touch the ground are: non-lead forelimb (NLFL) followed by lead forelimb (LFL) which propulses the body in a flight phase where the backlimbs are gathered under the body, then non-lead hindlimb (NLHL) contacts ground followed by lead hindlimb (LHL) which in turn propulses the body into the next aerial phase (modified from Hudson et al. 2012)

Optimized footfall patterns alone do not achieve the speeds observed in these animals, but the whole body structure, comprising tail and spine contribute to the movement. The cheetah's flexible vertebral column undergoes substantial flexion and extension, allowing an increase in stride length and acts as a spring, maximizing the return of energy in the movement (Bertram and Gutmann 2009). It also allows the animal to place the hind limbs directly under its center of mass, supporting 70% of the weight during gallop, while in most quadrupeds more than half of the body weight is supported by the forelimbs even during gallop. This capacity thought to enhance the grip during acceleration and maneuvering abilities (Hudson et al. 2012).

6.4 Bursts—Skipping, Hopping, Jumping

In *hopping* the two hind limbs move in phase so the body bounces in a spring-like motion, supported by the limbs. Mammals have a quite steady and continuous hopping gait, described as 'ricochetal' (Fig. 39).

Fig. 39 One sequence of a long jump in a man

Example	1. Wallabies (Fam. *Macropodidae*)

In wallabies, after the initial generation of force by the leg muscles, very little work is done to sustain the hopping: 60–96% of the energy is saved in the leg's elastic tendons and re-used in the next bounce. At 6 m/s metabolic energy expenditure is reduced by 50%, a female can for instance carry a pouch-young of 15% her own weight for free (Baudinette and Biewener 1998).

Catapult mechanism. The catapult muscle requires a preparation time to be drawn before the instant jump: the jumping muscle presses an extremely elastic cushion to draw the leg back. The propulsion muscle pulls then the corresponding tendon to release the stored energy in an instant jump.

Example	1. Fleas (or. *Siphonaptera*) jump up to 200 times their own height

Fleas are among the most successful jumpers in the animal kingdom. Such powerful and fast jumps exceed the direct muscle capacity: in the preparation phase of the jump, energy is stored in a resilin pad set under a torque force by the tendon of the primary jumping muscle passing behind the elastic joint; to release the catapulting mechanism, another muscle pulls the tendon in the opposite direction (Burrows 2009) (Fig. 40).

Kick mechanisms resemble the catapult mechanism as it also requires preparation and releases stored spring energy.

Example	2. Click beetles (fam. *Elateridae*)

On the front chest of the click-beetle sits a spike, which is usually locked into a small cavity. The beetle drags, through a powerful muscle, the front chest backwards until the spike rests on the edge of the cavity and then, unfolding great power, lets it slip into the original position. The quick nodding movement lifts both ends of the insect and bounces it into the air producing the "click" sound as the spike touches the cavity. See also deepened study in Chap. 6.

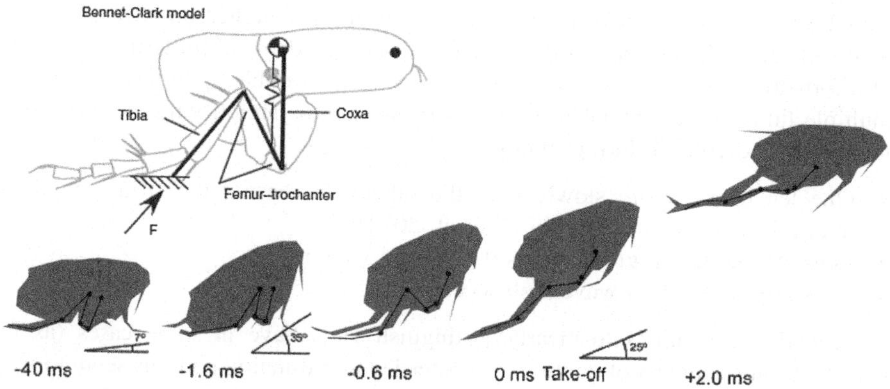

Fig. 40 Mechanical model and schematic jumping sequence of an adult hedgehog flea (sp. *Archaeopsyllus erinacei*) (based on Sutton and Burrows 2011)

Skipping is an uncommon gait adopted by bipeds for short bursts of locomotion. It has a similar footfall pattern as in gallop as it places successively the plantar in contact with the medium with a flight phase afterwards. Flap-bounding (see Flying locomotion) is an intermittent flight mode used by small birds, with similar characters to skipping.

Example	1. Humans (sp. *Homo sapiens*)

7 Multiple Limb Locomotion

A very broad range of animals, from ciliated systems to octopuses, can be included in the multiple limb category, as it groups all systems with more than four appendages. Despite the structural and environmental diversity of the systems here categorized (exoskeletal and hydrostatic, marine and terrestrial organisms), a number of reasons and common factors associate multiple limb systems, speaking for such a categorization in a biomimic purpose (Fig. 41).

Firstly, many locomotive patterns are unique to organisms with multiple appendages. In many cases, animals with many legs can move both in the direction of their head or sideways (Biewener 2007; Cruse and Graham 1985). A substantial difference with tetrapods, which all multiple limb systems share, is the growing

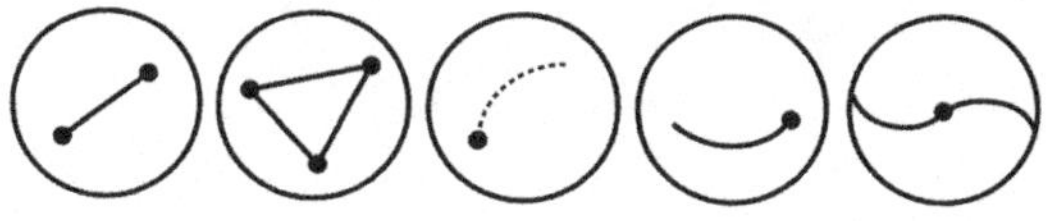

Fig. 41 Multiple limb gaits analyzed: bipedal, tripod, gallop, rowing and metachronal

complexity of inter-leg coordination and control as a higher number of legs is involved. To make up for an otherwise higher energy demand (in terms of amount of information to manage), and to avoid that the legs tumble upon their neighbors, multiple limbs are coordinated in groups or in patterns simplifying the amount of information to control during motion:

- One leg cycles more slowly than the others, increasing the stride length or letting it recover more slowly (Ji et al. 2014);
- Legs are moved in groups (tripod, tetrapod walking);
- Legs are moved in a wavelike succession.

Secondly, difficulties to clearly distinguish gaits have in some cases (as for insects) been the object of controversy. Inter-leg coordination patterns seem to vary depending on environmental conditions (as surface structure, slope, etc.), on speed (fast insects would use tripod walking, slow insects a metachronal wave, and a number of other patterns would occur in between the two extremes), and on load (Grabowska et al. 2012).

The following classification lists a number of recurring motion patterns, of which some are shared (with case specific variations) across systems with different numbers of appendages and across different media of locomotion, while other are specific to defined types of systems.

7.1 *Biped Gait*

Biped locomotion is rare in multiple-legged animals: it is a specific gait used only in few occasions and exclusively at high speeds, as the dynamic thrust allows for stability in body-types where standing on two legs is extremely difficult and statically unstable (Cruse et al. 2009) (Fig. 42).

Example	1. Ghost crab Ocypode (gen. *Ocypode*)

Fig. 42 A ghost crab, which runs on two legs at high speeds

The Ghost crab Ocypode (gen. *Ocypode*) becomes bipedal at its highest speeds. One theory is that the movement requires less coordination and allows to concentrate on the speed (Burrows and Hoyle 1973).

Example	2. Cockroaches (sp. *Periplaneta americana*)

The american cockroach (sp. *Periplaneta americana*) has been filmed running bipedally on its hind legs, incorporating an areal phase at each stride at around 1.0–1.5 m/s (Full and Tu 1991).

7.2 *Tripod Gait*

Tripod walking is one of the most typical insect gaits and is also one of the most stable variations of multiple legged locomotion, as support is never entirely lost throughout the stepping pattern. As three legs swing together, static equilibrium is achieved because the body's center of gravity falls within the triangular area of support represented by the limbs (Gruber 2011; Cruse et al. 2009) (Fig. 43).

Example	1. Ants walk resting alternatively on 3 legs per half-cycle: the front and the back leg on one side, the middle leg on the other side, inverting the sequence for the second half-cycle

At slower speeds, tripod gaits shift into what are commonly referred to as tetrapod gaits—where at least four legs are at the ground at any moment of time—although there are effectively no separate categories of gaits, but rather a continuum depending on walking speed (Cruse et al. 2009) and nature of the surface of support (inclination, roughness, etc.) (Grabowska et al. 2012). Although being different in appearance, both types of patterns move the contralateral leg pair out of phase (Smolka et al. 2013) (Fig. 44).

Example	1. Stick insects (sp. *Carausius morosus*) use a tetrapod gait for free-walking on horizontal surfaces (Grabowska et al. 2012)

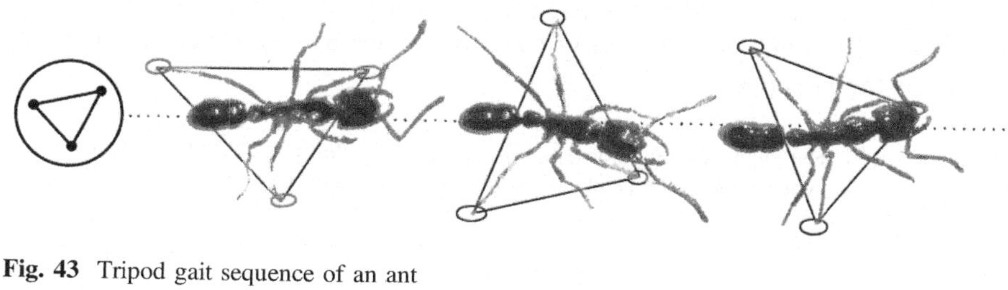

Fig. 43 Tripod gait sequence of an ant

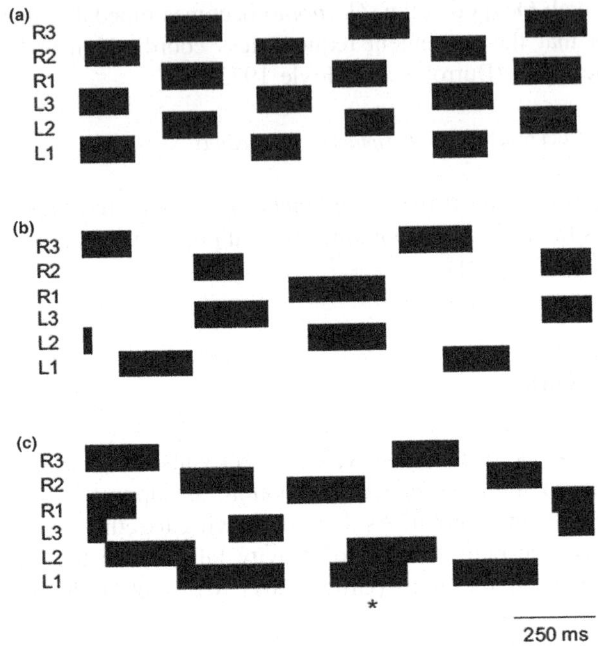

Fig. 44 Three stereotypic hexapedal walking patterns: (**a**) tripod gait, (**b**) and (**c**) tetrapod gaits (from Grabowska et al. 2012). Black bars indicate leg swing, not the footfall pattern

In octa-and deca-pods, different species use different number of legs for walking, using all available limbs or keeping some free for an eventual different use: Mictyris (gen. *Mictyris*) often walks on all 10, Astacus (gen. *Astacus*) usually uses 8 while rock lobsters Jasus (gen. *Jasus*) walk only on the posterior 3 legs.

Interlimb couplings are many and varied. Moreover some animals are adapted both for locomotion on land and in water. However all gaits need to provide propulsion and continuous support of the body, limiting the patterns of motion.

In the case of octopods, excluding the appendages with chelae, and because a walking cycle has no set starting point, there are six possible gaits: 2345, 2354, 2435, 2453, 2534 and 2543 (patterns 3452, 4523 and 5234 are the same as 2345).

In the case of decapod crustaceans, studies have shown that there are no abrupt changes in gait to increase or decrease in speed (Clarac 1985).

A variation of the tripod gait is also used in some jumping and bouncing actions. Hexapods use a similar gait observed in vertebrates, a bouncing gait alternating phases of tripod support.

Example	1. Harpagenathos saltator (sp. *Harpagenathos saltator*) is able to jump by synchronising the action of its mid and hind pairs of legs

7.3 Galloping Gait

Galloping gaits are quite rare in most insect species, as asynchronous stepping patterns as the tripod gait are the most used, with the exception of uncommon circumstances, as when swimming or stepping over an obstacle (Smolka et al. 2013) (Fig. 45).

Example	1. Desert dung beetle (sp. *Pachysoma endroedyi*)

This beetle uses its hind limbs to drag dung and debris towards its burrow, synchronously alternating both front and both mid-limbs to move. The hind limbs do not appear to contribute to the propulsion, even when no load is carried. Although no aerial phase occurs, as either fore or mid limbs is in contact with the ground, the leg coordination can be compared to the bouncing gallop of a rabbit. This type of gait appears to be slower than the standard tripod gait (Smolka et al. 2013).

Example	2. Fishing spider (sp. *Dolomedes triton*)

These spiders are known for hunting and walking on the water surface. For moving at faster speeds, the spiders move the anterior three pairs of legs synchronously, sweeping them downwards and backwards in the thrust phase, follows an aerial phase where the limbs have little or no contact with the water surface and the limbs return forwards, preparing for the next stride (Suter and Wildman 1999) (Fig. 46).

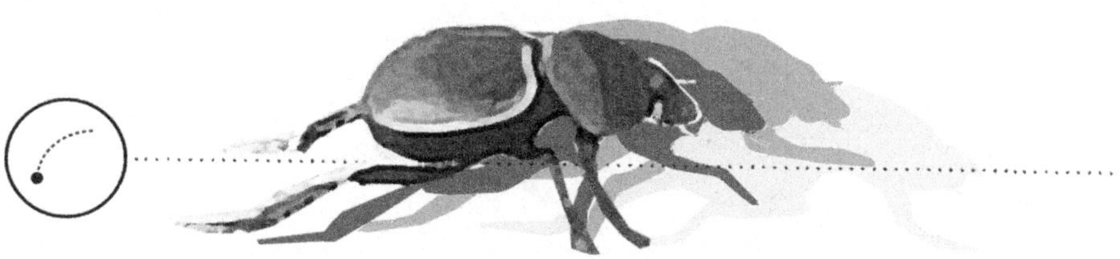

Fig. 45 A desert dung beetle in a galloping sequence

Fig. 46 Galloping sequence in a fishing spider, ellipses indicate contact with the water surface (from Suter and Wildman 1999)

7.4 Rowing Locomotion

Rowing is used by animals to drag themselves, or glide, forwards by achieving a horizontal thrust on the substrate (Fig. 47).

Example	1. Brittle star (sp. *Ophiuroidae echinata*)

Brittle stars are an interesting case study of a pentaradially symmetrical organism using the bilateral symmetrical coordinated action of its five multi-jointed limbs to propel it forward on the substrate. One chosen limb is directed forwards (rowing) or backwards (reverse rowing), while the two front limbs propel the body and the hind limbs are dragged passively, or are used to explore the substrate. Rowing locomotion in the brittle star can be compared to the terrestrial locomotion of sea turtles (Astley 2012) (Fig. 48).

To change direction of locomotion, the brittle star keeps the same footfall pattern and chooses a new lead limb, as the other arms take on the subordinate roles (Astley 2012).

Example	2. Fishing spider (sp. *Dolomedes triton*)

Fishing spiders use rowing as slow and medium speed locomotion when hunting on the water surface. The two middle pairs of the legs are swung forwards and pushed backwards on the water surface, allowing the spider to glide over the surface while supported by the front and the back pairs of limbs (Suter and Wildman 1999).

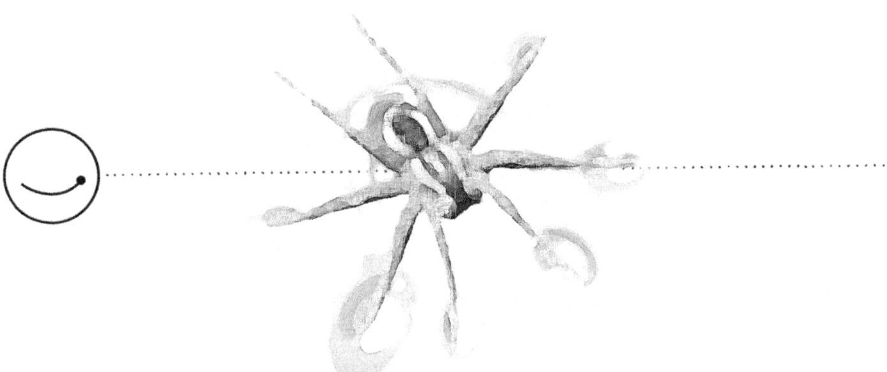

Fig. 47 A fishing spider waiting for a prey on the water surface

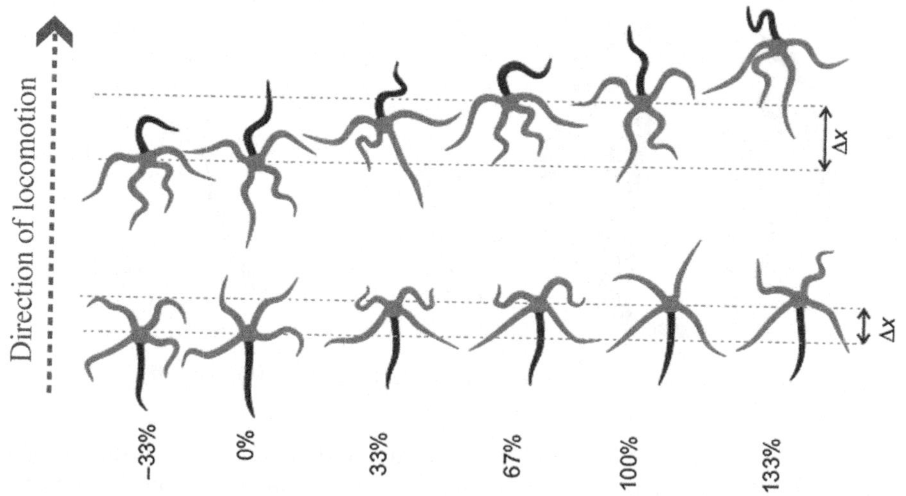

Fig. 48 Rowing (above), and reverse rowing (below), sequences in a brittle star (adapted from Astley 2012) can be identified by the center limb, colored black. The percentages express the timing in the locomotion cycle, and Dx the distance traveled during the cycle

7.5 Metachronal Wave

Metachronal rhythm is a sequential wave-like activity moving a multitude of component structures together, to maintain a constant phase relationship between adjacent members, so they do not to bump into each other. It enhances the efficiency reducing the occurrence of interference between the parts. Most walking and all swimming systems have developed variations of this physiological mechanism. In symplectic metachronal rhythm the wave moves from anterior towards posterior, in antiplectic from posterior to anterior (Sleigh and Barlow 1980) (Fig. 49).

In multiple limb systems, the feet are moved successively in a wave-like order providing forward propulsion. The advantage of developing a multitude of feet and to use a wave like motion is to move over an uneven terrain by providing punctual support.

Example	1. Centipedes, millipedes

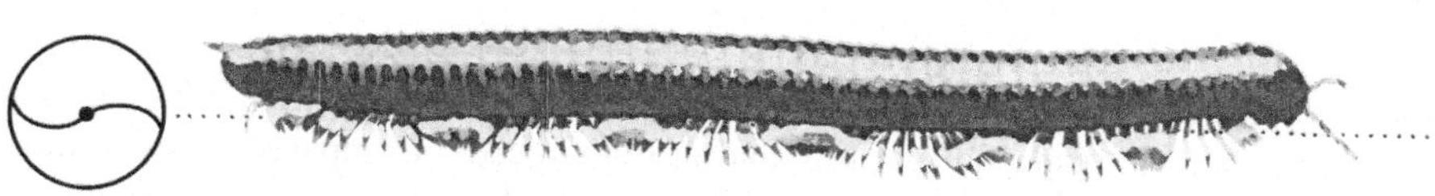

Fig. 49 Metachronal walking motion of a millipede

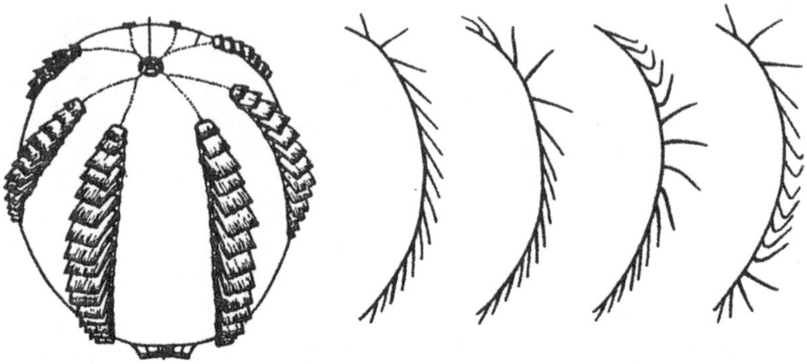

Fig. 50 Sequence of ciliary locomotion (modified from Elder and Trueman 1980)

Ciliary locomotion makes use of thousands of short (15 μm) hair-like appendages (cilia) located along the surface of a cell, which cannot be properly compared to limbs, but certainly fulfill a similar function (Biewener 2007). As ciliary swimming exists only in micro-organisms, where hydrodynamic efficiency is predicted to decrease with increasing size (the upper limit to ciliary body size is around 0,1 mm), the emergence of a metachronal wave is essential for an efficient propulsion. While the effect of a single cilium is negligible, a synchronized beating of cilia can achieve a traveling wave at the end of the tips (Ghorbani and Najafi 2017) (Fig. 50).

Example	1. Locomotion mode: Volvox (Gen. *Volvox*); 2. Locomotion mode: Paramecium (Gen. *Paramecium*); 3. Mucociliary transport in respiratory systems: vertebrates (Ghorbani and Najafi 2017)

The propulsion involves synchronized pulsations of the cilia beating with a slight phase difference to generate a symplectic and antiplectic metachronal waves—parallel and antiparallel to the direction of the power stroke respectively (Ghorbani and Najafi 2017). During each beat, the cilia bends from the base and propagates the bending movement towards the tip, achieving an effective and a recovery stroke generating thrust (Ishikawa 2009).

8 Flying Locomotion

Flapping flight is a highly energy-requiring mean of locomotion with extreme metabolic demand, especially at low and at high speeds. Compared to swimming, where buoyancy supports the body weight and only drag works against propulsion, flying requires constant power to support the body in the air and provide thrust. Flight is however a very versatile and fast mean of locomotion. The overall Cost of

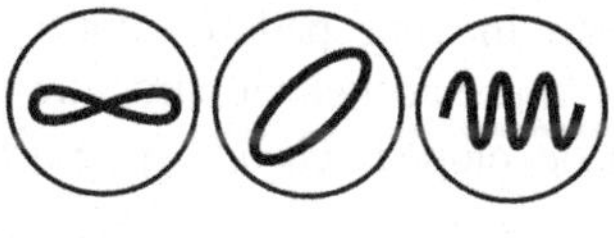

Fig. 51 Three types of flying locomotion (from the left): eight-shaped wing stroked flight, elliptical wing stroked flight, flap bounding

Transport is therefore very efficient and relatively low compared with terrestrial locomotion. Moreover, capacity to invade a new medium opens up ecological niches with adequate food supply and good security from predation, hence the strong selective pressure towards development of capacity to generate lift to move through the air (Rayner 1981) (Fig. 51).

As power supply is generally a main a limiting parameter, most adaptations to flying locomotion address the issue of cutting energy consumption through a low weight of the animal, higher efficiency in the wing beat patterns and maximum use of available latent Energy sources contributing to the locomotion. Any flight adaptation is achieved through a compromise between a broad range of ecological, physiological, behavioral and mechanical demands. It specifically requires an anatomical surface-area to support the animal's weight by using airfoil action[1] and generating lift (Biewener 2007). This is achieved through:

(1) Adaptation of the limbs, which are either specialized for flight (providing thrust through flapping flight) or adapted for another medium and are secondarily used for gliding, while thrust is provided in another way (see passive locomotion).

Example	1. Birds, forelimbs are used exclusively for flight 2. Gliding tree frog (sp. Agalychnis spurrelli) have webbed feet which allow it to glide after providing initial thrust though jumping 3. Flying fishes jump out of the water and glide with their fins

(2) Development of membranes along the body.

Example	1. Flying dragons (gen. Draco) are lizards which have developed an aerofoil membrane by lengthening their ribs 2. Flying geckos, squirrels and bats have all evolved a membrane connecting their limbs to the trunk

Control of the flight surface. As the airfoil might not be always needed, the design must be adaptable and flexible during other types of locomotion as well as during flight. The airfoil must be rigid during flight and easy to be rapidly moved to

[1]The cross-section of the wing modifies the air flow, generating a pressure difference between the upper and the lower side of the wing. Lift supports the body acting perpendicularly to the airflow, while drag retards the body acting parallel to the airflow. The angle of incidence of the airfoil affects the lifting effect: if the angle is too large, the pressure difference falls causing the airfoil to stall. Natural wings are well adapted to prevent stalling and undesired turbulences (Rayner 1981).

counter sudden crisis, change moving patterns during take-off and landing. This complex system requires a musculoskeletal structure apt to tolerate the stresses involved in the flight and a developed sensory system to obtain and interpret information while flying.

Size and shape of the wing surface determines, among other factors, the speed and the maneuverability of the flight. More lift can be generated either by flying faster or with longer wings (Rayner 1981). Wing design and flight patterns are strongly connected to the habitat and the foraging habits. Insectivores are mostly active flyers, while those feeding on the ground or on trees need less maneuverability. To describe the wings' proportions to the rest of the animal's geometry, two parameters are used:

- Wing load is the weight supported by unit area of the membrane. Slow flying birds generally have large wings with low wing loads to produce enough lift. Higher wing loads give the animal more gravitational potential energy and are distinctive of fast flyers.
- Aspect ratio (AR) is the wingspan to the wing chord. It describes the geometrical shape of the wing. The larger the surface, the bigger the lift and the smaller the rate of descent resulting in a good gliding performance. The most of the Energy cost in gliders goes for takeoff because of the difficulty of flapping with large wings. On the other hand smaller surfaces are very maneuverable but require more flapping (Norberg 1981).

Example	1. Albatrosses (fam. *Diomedeidae*) are gliders with large wingspan; 2. Swallows (fam. *Hirundinidae*) have small and manoeuvrable wings for quick and sudden moves

8.1 Eight-Shaped Wing Stroked Flight

In this wing beat cycle both down stroke and upstroke produce lift and thrust: flying insects and hummingbirds have evolved a unique shoulder articulation which allows them to rotate the wing to reverse the rotation in a typical 8-pattern, achieving a positive angle of attack during the upstroke (Brodsky 1994) (Fig. 52).

Hovering involves that the animal remains suspended in the air without moving forward. The wings oscillate in an almost horizontal plane producing lift and no thrust. To change and fly forwards, the animal inclines the stroke angle of its wings to produce thrust as well (Fig. 53).

Example	1. Bumblebees (gen. *Bombus*) 2. Hummingbirds (fam. *Trochilidae*)

Fig. 52 Initial stage of the backstroke in a hovering hummingbird, where the wing is turned upside-down

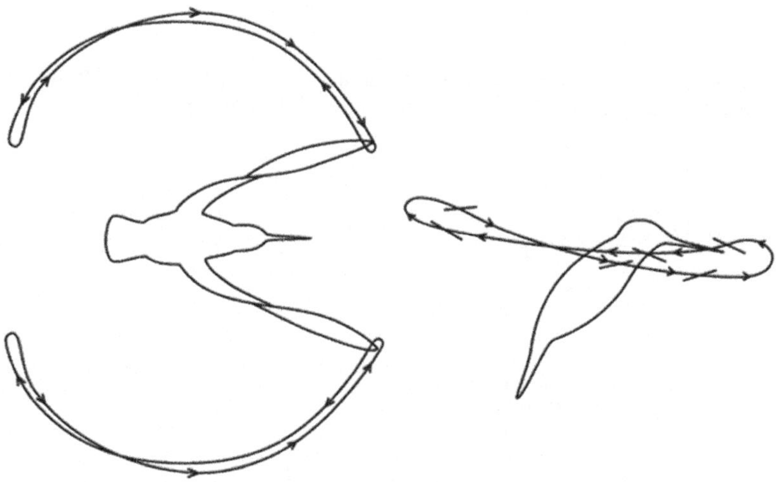

Fig. 53 Eight-shaped wing stroke typical of hovering in a hummingbird, seen from the side and from the top (modified from Videler 2005)

8.2 Elliptical Wing Stroked Flight

The wing beat cycle is basically divided into down stroke (producing lift and thrust) and upstroke bringing the wing to the initial position. Changes in wing area occur during each phase of a wing beat cycle: during upstroke the wing is collapsed or to reduce drag, during down stroke the wing is fully extended to maximize wing area.

In vertebrates the down stroke produces most of the lift and thrust required for flight (Biewener 2007) (Figs. 54, 55 and 56).

Example	1. Pigeons and doves (fam. *Columbidae*) 2. Eagles (fam. *Accipitridae*)

Fig. 54 Back stroke in a flying pigeon. The wing area changes geometry during up and down stroke to minimize drag in the first case and maximize thrust in the second

Fig. 55 Sequence of a pigeon taking off, showing the complex shape changes of the wing (modified from Zuk and Clark 1970)

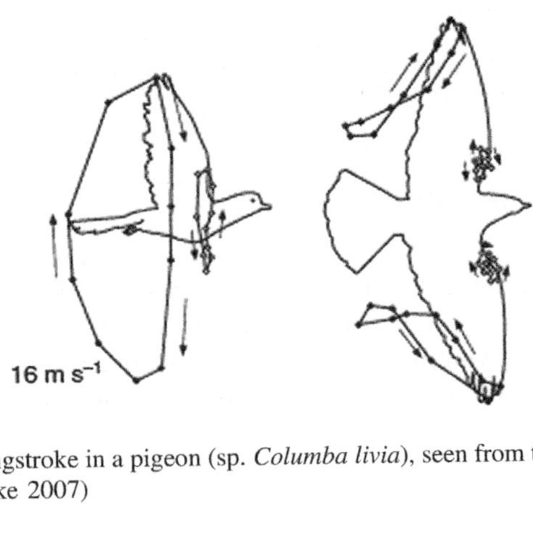

Fig. 56 Elliptical wingstroke in a pigeon (sp. *Columba livia*), seen from the side and from the top (adapted from Tobalske 2007)

8.3 Flap-Bounding

Flap-bounding is an intermittent flight mode where small birds fly up and down in a sinusoidal path, alternating fast flapping movements with short bounding periods propelling the bird upwards, with the wings folded against the body. Small-sized birds adopt this irregular flight mode, supposedly to reduce drag by improving the aerodynamics, thus saving energy (Videler 2005; Tobalske et al. 1999) (Figs. 57 and 58).

Example	1. Zebra finch (sp. *Taeniopygia guttata*)

Fig. 57 A zebra finch in the flapping flight and in the bounding phase

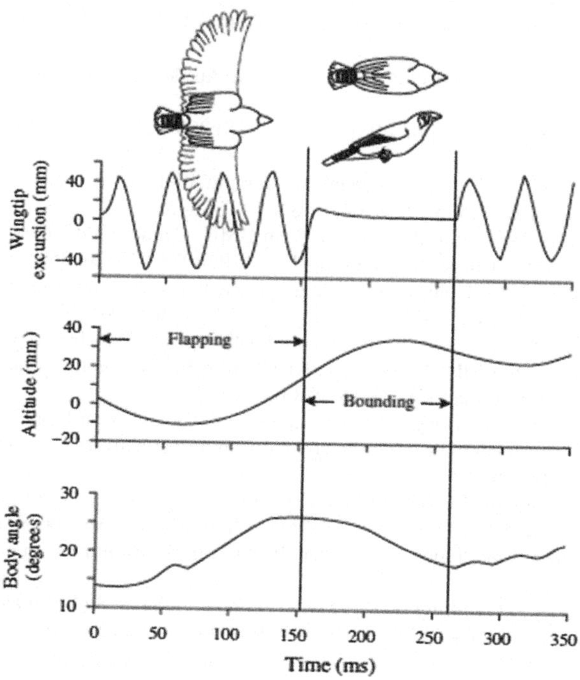

Fig. 58 Flap-bounding in a zebra finch flying at 12 m/s in a wind tunnel (from Videler 2005)

References

Alexander RMcN (1981) The gait of tetrapods: adaptations for stability and economy. In: Day MH (ed) Vertebrate locomotion, symposia of the Zoological Society of London, nr.48. Academic Press Inc., London, UK. ISBN 0-12-613348-4

Alexander RMcN (2000) Storage and release of elastic energy. In Nigg BM, MacIntosh BR, Mester J (eds) Biomechanics and biology of movement. Human kinetics. Champaign, Ill. ISBN 0-7360-0331-2

Alexander RMcN (2004) Bipedal animals, and their differences from humans. J Anat 204:321–330. https://doi.org/10.1111/j.0021-8782.2004.00289.x

Alexander RMcN (2005) Models and the scaling of energy costs for locomotion. J Exp Biol 208:1645–1652. https://doi.org/10.1242/jeb.01484

Altman JS, Kien J (1985) The anatomical basis for intersegmental and bilateral coordination in locusts. In: Bush BMH, Clarac F (eds) Coordination of motor behaviour. Cambridge University Press, Cambridge UK. ISBN 978-0521264259

Astley HC (2012) Getting around when you're round: quantitative analysis of the locomotion of the blunt-spined brittle star, *Ophiocoma echinata*. J Exp Biol 215:1923–1929. https://doi.org/10.1242/jeb.068460

Baudinette RV, Biewener A (1998) Young wallabies get a free ride. Nature 395:653–654. https://doi.org/10.1038/27111

Bertram JEA, Gutmann A (2009) Motions of the running horse and cheetah revisited: fundamental mechanics of the transverse and rotary gallop. J R Soc Interface 6(35):549–559. https://doi.org/10.1098/rsif.2008.0328

Biewener AA (2006) Patterns of mechanical energy change in tetrapod gait: pendula, springs and work. J Exp Zool Part A, 899–911. https://doi.org/10.1002/jez.a.334

Biewener AA (2007) Animal locomotion. Oxford University Press, Oxford, UK. ISBN 978-0-19-850022-3

Brodsky AK (1994) The evolution of insect flight. Oxford University Press, Oxford, UK. ISBN 0-19-854681-5

Brown ME (2013) The physiology of fishes: behavior. Academic Press. Inc., New York. ISBN 9781483262673

Burrows M (2009) How fleas jump. J Exp Biol 212(18):2881–2883. https://doi.org/10.1242/jeb.022855

Burrows M, Hoyle G (1973) The mechanism of rapid running in the ghost crab, *Ocypode ceratophthalma*. J Exp Biol 58:327–349. ISSN 1477-9145

Bush BMH, Clarac F (1985) Introduction. In: Bush BMH, Clarac F (eds) Coordination of motor behaviour. Cambridge University Press, pp 1–7. ISBN 978-0521264259

Butler SM, Camilli A (2005) Going against the grain: chemotaxis and infection in Vibrio cholerae. Nat Rev Microbiol 3(8):611–620. https://doi.org/10.1038/nrmicro1207

Clarac F (1985) Peripheral influences on the coordination of the legs during walking in decapod crustaceans. In Bush BMH, Clarac F (eds) Coordination of motor behaviour. Cambridge University Press, Cambridge UK. ISBN 978-0521264259

Cruse H, Graham D (1985) Models for the analysis of walking in arthropods. In Bush BMH, Clarac F (eds) Coordination of motor behaviour. Cambridge University Press, Cambridge UK. ISBN 978-0521264259

Cruse H, Durr V, Schilling M, Schmitz J (2009) Principles of insect locomotion. In Arena P, Patanè L (eds) Spatial temporal patterns for action-oriented perception in roving robots. Cognitive systems monographs 1. Springer, Berlin, pp 43–96. https://doi.org/10.1007/978-3-540-88464-4_2

Dickinson MH, Farley CT, Full RJ, Koehl MAR, Kram R, Lehman S (2000) How animals move: an integrative view. Science 288(5463):100–106. https://doi.org/10.1126/science.288.5463.100

Day MH (ed) (1981) Vertebrate locomotion, symposia of the Zoological Society of London no.48, The Zoological Society of London, Academic Press Inc., Oxford, London. ISBN 9780126133486

Elder HY, Trueman ER (eds) (1980) Aspects of animal movement. Cambridge University Press, Cambridge, USA. https://doi.org/10.1113/expphysiol.1981.sp002536

Full RJ, Tu MS (1991) Mechanics of a rapid running insect: two-, four-, and six-legged locomotion. J Exp Biol 156:215-231. ISSN 1477-9145

Gans C (1985) Motor coordination factors in the transition from tetrapody to limblessness in lower vertebrates. In: Bush BMH, Clarac F (eds) Coordination of motor behaviour. Cambridge University Press, pp 183–199. ISBN 978-0521264259

Gans C (1986) Locomotion of limbless vertebrates: pattern and evolution. Herpetologica 42 (1):33–46

Ghorbani A, Najafi A (2017) Symplectic and antipectic waves in an array of beating cilia attached to a closed body. American Physical Society, Phys Rev E 95:5. https://doi.org/10.1103/physreve.95.052412

Goldspink G (1981) The use of muscles during flying, swimming, and running from the point of view of energy saving. In Day MH (ed) Vertebrate locomotion, symposia of the Zoological Society of London, nr.48. Academic Press Inc., London, UK. ISBN 0-12-613348-4

Grabowska M, Godlewska E, Schmidt J, Daun-Gruhn S (2012) Quadrupedal gaits in hexapod animals – inter-leg coordination in free-walking adult stick insects. J Exp Biol 215:4255–4266. https://doi.org/10.1242/jeb.073643

Gruber P (2011) Biomimetics in architecture, architecture of life and buildings. Springer-Verlag, Wien. ISBN 978-3-7091-0331-9

Guo ZV, Mahadevan L (2008) Limbless undulatory propulsion on land. PNAS 105(9):3179–3184. https://doi.org/10.1073/pnas.0705442105

Hale ME, Day RD, Thorsen DH, Westneat MW (2006) Pectoral fin coordination and gait transitions in steadily swimming juvenile reef fishes. J Exp Biol 209:3708–3718. https://doi.org/10.1242/jeb.02449

Harischandra N, Knuesel J, Kozlov A, Bicanski A, Cabelguen J-M, Ijspeert A, Ekeberg O (2011) Sensory feedback plays a significant role in generating walking gait and in gait transition in salamanders: a simulation study. Front Neuro 5(3):1–13. https://doi.org/10.3389/fnbot.2011.00003

Hildebrand M (1989) The quadrupedal gaits of vertebrates. BioScience 39(11), Animals in Motion 766–775. https://doi.org/10.2307/1311182

Hoyt DF, Taylor CR (1981) Gait and energetics of locomotion in horses. Nature 292:239–240. https://doi.org/10.1038/292239a0

Hudson PE, Corr SA, Wilson AM (2012) High speed galloping in the cheetah (*Acinonyx jubatus*) and the racing greyhound (*Canis familiaris*): spatio-temporal and kinetic characteristics. J Exp Biol 215:2425–2434. https://doi.org/10.1242/jeb.066720

Ishikawa T (2009) Suspension biomechanics of swimming microbes. J R Soc Interf 6:815–834. https://doi.org/10.1098/rsif.2009.0223

Ji A, Lei Y, Wang J, Ni Y, Dong B, Dai Z (2014) Variation in spatial and temporal kinematics of level, vertical and inverted locomotion on a stinkbug *Erthesina fullo*. Chin Sci Bull 59 (26):3333–3340. https://doi.org/10.1007/s11434-014-0310-x

Kier WR (2012) The diversity of hydrostatic skeletons. J Exp Biol 215:1247–1257. https://doi.org/10.1242/jeb.056549

Lees J, Gardiner J, Usherwood J, Nudds R (2016) Locomotor preferences in terrestrial vertebrates: an online crowd sourcing approach to data collection. Scientific Reports 6 (28825), Nature. https://doi.org/10.1038/srep28825

Leman M, Moelants D, Varewyck M, Styns F, van Noorden L (2013) Activating and relaxing music entrains the speed of beat synchronized walking. PLoS ONE 8(7). https://doi.org/10.1371/journal.pone.0067932

Lindsay CC (1978) Form, function and locomotory habits in fish. In: Hoar D, Randall DJ (eds) Fish physiology, vol VII. New York Academic Press, pp 9–10

Lissmann HW (1949) Rectilinear locomotion in a snake (Boa occidentalis). J Exp Biol 26:368–379. ISSN 1477-9145
Marey EJ (1884) Geometric chronophotograph of a man in a black suit. In: Jensenius AR (ed) Some video abstraction techniques for displaying body movement in analysis and performance. Leonardo: J Int Soc Arts, Sci Technol 46(1):53–60 (2013)
Marvi H, Bridges J, Hu DL (2013) Snakes mimic earthworms: propulsion using rectilinear travelling waves. J Roy Soc Interf 10. https://doi.org/10.1098/rsif.2013.0188
Massion J (1985) Postural changes during movement performance in a tetrapod. In Bush BMH, Clarac F (eds) Coordination of motor behaviour. Cambridge University Press, Cambridge UK. ISBN 978-0521264259
Miller S, Schomburg ED (1985) Locomotor coordination in the cat. In: Bush BMH, Clarac F (eds) Coordination of motor behaviour. Cambridge University Press, Cambridge UK. ISBN 978-0521264259
Minetti AE (2000) The tree modes of terrestrial locomotion. In Nigg BM, MacIntosh BR, Mester J (eds) Biomechanics and biology of movement. Human kinetics. Champaign, Ill. ISBN 0-7360-0331-2
Moon B (2001) Snake locomotion. University of Louisiana at Lafayette. Accessed 15 February 2018 from https://www.ucs.louisiana.edu/~brm2286/locomotn.htm
Muramatsu K, Yamamoto J, Abe T, Sekiguchi K, Hoshi N, Sakurai Y (2013) Oceanic squid do fly. Marine biology, Springer. https://doi.org/10.1007/s00227-013-2169-9
Nombela C, Hughes LE, Owen AM, Grahn JA (2013) Into the groove: can rhythm influence Parkinson's disease? Neurosci Biobehav Rev 37:2564–2570. https://doi.org/10.1016/j.neubiorev.2013.08.003
Norberg UM (1981) Flight, morphology and the ecological niche in some birds and bats. In Day MH (ed) Vertebrate locomotion, symposia of the Zoological Society of London, nr.48, Academic Press Inc., London, UK. ISBN 0-12-613348-4
O'Connor SM, Dawson TJ, Kram R, Donelan JM (eds) (2014) The kangaroo's tail propels and powers pentapedal locomotion. Biol Lett 10. Royal Society Publishing. https://doi.org/10.1098/rsbl.2014.0381
Rayner JMV (1981) Flight adaptations in vertebrates. In: Day MH (ed) Vertebrate locomotion, symposia of the Zoological Society of London, nr.48, Academic Press Inc., London, UK. ISBN 0-12-613348-4
Richardson MJ, Marsh KL, Isenhower RW, Goodman JRL, Schmidt RC (2007) Rocketing together: dynamics of intentional and unintentional interpersonal coordination. Hum Mov Sci 26(6):867–891. https://doi.org/10.1016/j.humov.2007.07.002
Robert BL (1981) The organization of the nervous system of fishes in relation to locomotion. In Day MH (ed) Vertebrate locomotion, symposia of the Zoological Society of London, nr.48, Academic Press Inc., London, UK. ISBN 0-12-613348-4
Roerdink M, Daffertshofer A, Marmelat V, Beek PJ (2015) How to sync to the beat of a persistent fractal metronome without falling off the treadmill? PLoS ONE 10(7). https://doi.org/10.1371/journal.pone.0134148
Rubenson J, Heliams DB, Lloyd DG, Fournier PA (2004) Gait selection in the ostrich: mechanical and metabolic characteristics of walking and running with and without aerial phase. Proc R Soc Lond B 271:1091–1099. https://doi.org/10.1098/rspb.2004.2702
Shiba K, Shibata D, Inaba K (2014) Autonomous changes in the swimmimg direction of sperm in the gastropod *Strombus luhanus*. J Exp Biol 217:986–996. https://doi.org/10.1242/jeb.095398
Seymour MK (1969) Locomotion and coelomic pressure in Lumbricus Terrestris L. J Exp Biol 51:47–58. ISSN 1477-9145
Sleigh MA, Barlow DI (1980) Metachronism and control of locomotion in animals with many propulsive structures. In: Elder HY, Trueman ER (eds) Aspects of animal movement. Cambridge University Press. https://doi.org/10.1113/expphysiol.1981.sp002536
Smolka J, Bryne MJ, Scholtz CH, Dacke M (2013) A new galloping agit in an insect. Curr Biol 23 (20):R913–R915. https://doi.org/10.1016/j.cub.2013.09.031

Soffe SR (1985) Central coordination of swimming in lower vertebrates. In: Bush BMH, Clarac F (eds) Coordination of motor behaviour. Cambridge University Press, pp 141–159. ISBN 978-0521264259

Suter RB, Wildman H (1999) Locomotion on the water surface: hydrodynamic constraints on rowing velocity require a gait change. J Exp Biol 202:2771–2785. ISSN 1477-9145

Sutton PG, Burrows M (2011) Biomechanics of jumping in the flea. J Exp Biol 214:836–847. https://doi.org/10.1242/jeb.052399

Tobalske BW (2007) Biomechanics of bird flight. J Exp Biol 210:3135–3146. https://doi.org/10.1242/jeb.000273

Tobalske BW, Peacock WL, Dial KP (1999) Kinematics of flap-bounding flight in the zebra finch over a wide range of speeds. J Exp Biol 2020:1725–1739. ISSN 1477-9145

Tunca C, Pehlivan N, Ak N, Arnrich B, Salur G, Ersoy C (2017) Inertial sensor-based Robust Gait analysis in non-hospital settings for neurological disorders. Sensors 17(825). https://doi.org/10.3390/s17040825

Valentine JW (2004) On the origin of phyla. The University of Chicago Press, Chicago and London. ISBN 0-226-84548-6

Videler JJ (2005) Avian flight. Oxford University Press, Oxford, UK. ISBN 978-0-19856603-8

Von Holst E (1973) The behavioural physiology of animals: the selected papers of Erich von Holst. University of Miami Press, ISBN, p 9780870242618

Webb GJW, Gans C (1982) Galloping in crocodylus johnstoni—a reflection of terrestrial activity? Records of the Australian Museum 34(14):607–618. https://doi.org/10.3853/j.0067-1975.34.1982.244

Wilson AM, Van den Bogert AJ, McGuian MP (2000) Optimization of the muscle-tendon unit for economical locomotion in cursorial animals. In: Herzog W (ed). Skeletal muscle mechanics, from mechanisms to function. John Wiley & Sons, Ltd., Chichester, UK, pp 517–545. ISBN: 978-0-471-49238-2

Zug GR (2017) Locomotion. Encyclopædia Britannica, inc. Accessed 20 February 2018 from https://www.britannica.com/topic/locomotion/Anguilliform-locomotion

Zuk W, Clark RH (1970) Kinetic architecture. Van Nostrand Reinhold, New York. ISBN 0442156723

Chapter 6
Case Studies

Abstract The following part aims to deepen some of the aspects discussed in the previous chapters by analyzing fifteen systems in nature and their specific characters of motion. These case studies combine the previously discussed parameters with smart design, and have been chosen from three main categories of systems with rising complexity in the control of the movements: plants, invertebrates and vertebrates.

Plant movements, and especially the mechanics of seed dispersal, are designed with simple structures and kinetics, compared to the complexity of animal locomotion. The absence of neural networks (or the presence of more basic ones), the environmentally triggered mechanisms and the subsequent chain reactions triggering motion, are three main aspects that make plants extremely interesting for a biomimetic technological transfer.

Types of movements used by animals, both in locomotion and for a wide range of movements are divided into mechanics of soft-bodied systems and mechanics of rigid systems.

Each case study is characterized by many of the features analyzed in the previous chapters. A simplified summary of the main combinations of geometrical and mechanic characters in these animal and plants can be visualized at the end of Chaps. 3 and 4.

1 Plants and Seeds

Plant movements are mostly based upon an exchange in fluids within the plant—swelling up upon liquid intake and shrinking when undergoing water loss—regulating the internal pressure (turgor pressure). Most tropic movements are carried out based on this type of mechanics, as the sunflower following the sun path, leaves opening and closing for day/nighttime. Plant cell's pressure can be regulated actively (controlled by the plant), or passively (controlled by the environmental conditions, as drought) (Del Campo and Whitaker 2008).

S. Persiani, *Biomimetics of Motion*,
https://doi.org/10.1007/978-3-319-93079-4_6

All kinds of movements in plants can be placed into two main categories.

a. Slow hydraulic movements;
b. Rapid movements due to the release of built-up elastic potential energy.

Rapid motion in plants is based on the release of stored elastic energy and is mainly used for seed and sporangium dissemination, defense and feeding. Among the techniques used in rapid movement, two sub-categories can be identified:

- Snap-buckling geometries;
- Explosive and fracture dominated geometries.

The advantage with the former one is the reversibility of the mechanism, which switches between two opposite and complementary shapes, not involving the tearing of the plant tissue to release the stored energy. On the other hand, the fracture-dominated mechanisms are generally also the fastest ones (Del Campo and Whitaker 2008). Most often, these two kinds of plant movement can be found to work together, completing each other: the hydraulic mechanisms build up energy in the structures, bringing them to the limit of structural stability, and the fracture or geometrical snap-buckling instantly releases the energy.

The case-study sheets explore the following vegetal systems, with a specific focus on one of its kinetic features:

Name of the organism	Kinetic aspect studied
Bulliform cells	Hydraulic mechanics
Himalayan balsam	Seed dispersal mechanics
Fern	Spore dispersal mechanics
Horsetail	Spore dispersal mechanics
Morning glory	Photonastic flower opening
Mimosa pudica	Thigmonastic defense
Venus flytrap	Thigmonastic hunting

1.1 Bulliform Cells, Cell Mechanics in Rice Leaves

The physical shape of leaves in grass-like plants constantly adapts to the surrounding environment by changing shape. Basic leaf functions as transpiration, respiration and photosynthesis strongly depend on the geometry assumed by the surface of the leaf (Zou et al. 2011). The elongated ribbon-like flat leaf surfaces are primarily responsible for photosynthesis (Rost 1997), and among the studies on leaf mechanics, the osmotic adjustment of crops as rice during drought periods (specifically *Oryza sativa L.*) are among the most studied (Fig. 1).

Motion patterns. Leaves roll up during periods of drought to minimize transpiration. This shape is thought to help the plants to reduce water losses by

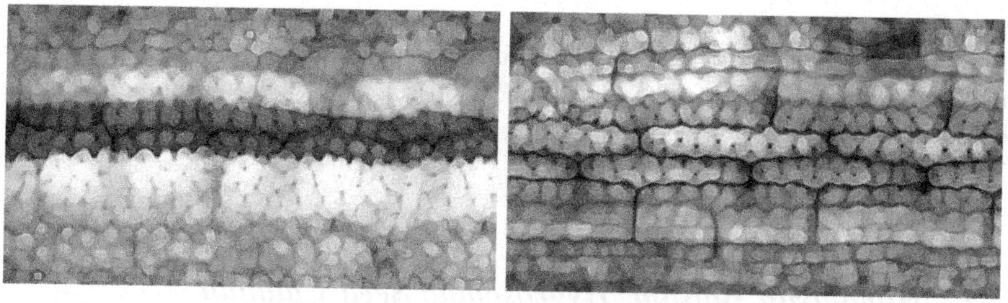

Fig. 1 Shrunken (left) and swollen (right) bulliform cells within a grass leaf

minimizing the exposure of the surfaces, which however also reduces the potential for photosynthesis. The rolling is therefore delayed to when water levels becomes critically low, with priority over energy-foraging (Hsiao et al. 1984).

Two different types of leaf transverse rolling occur in plants (Xiang et al. 2012):

a. Inward rolling (adaxial), depends on environmental factors as water deficiency, high air temperature, sunlight;
b. Outward rolling (abaxial).

Motion mechanism. Turgor pressure is a hydrostatic pressure generated in plant cells and responsible for growth, transport, cell metabolism and movement. Plants work like hydraulic machines taking in and releasing water through the semipermeable membranes of their cells. Regulation of the turgor pressure is how the plant achieves physical shape change in their tissues, which is determined by intracellular pressure and cell wall loosening in adaptation to the changing environment (Pritchard 2001; Routier-Kierzkowska et al. 2012) (Fig. 2).

The motor cells responsible for the shape changes are the 'bulliform cells', which can be pictured as large and thin enclosed compartments filled with water (Zou et al. 2011). These occur in groups in mechanically strategic points of the leaf:

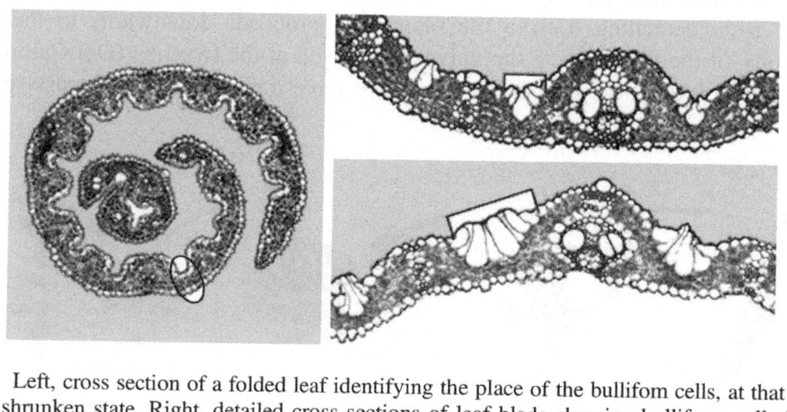

Fig. 2 Left, cross section of a folded leaf identifying the place of the bullifom cells, at that point in the shrunken state. Right, detailed cross sections of leaf blade showing bulliform cells in the shrunken (top) and the swollen state (bottom) (modified from Zou et al. 2011)

between vascular bundles on the upper side of the leaf's blade. When the surrounding atmosphere dries up, the bulliform cells loose turgor pressure and shrink, making the upper side of the leaf contract and roll up. When the humidity in the environment rises again, the cells expand opening up the leaves again (Xiang et al. 2012).

1.2 Himalayan Balsam, Hygroscopic Seed Catapult

The Himalayan balsam (*Impatiens glandulifera*) is an annual bush-like plant belonging to the species Impatiens, meaning "impatient" in Latin, because of the tendency of the species' mature seed capsules to explode when touched (Clements et al. 2008), firing the seeds up to 6 m away from the parent plant at speeds up to 6 m/s (ARKive 2016; Chapman and Gray 2012; Deegan 2012). It seems that the *I. glandulifera* has evolved different ballistic trajectories depending on the wind conditions to increase the distance of dispersal (Chapman and Gray 2012) (Fig. 3).

This type of seeds are built to minimize air friction, and are therefore heavier and more aerodynamic than wind-dispersed seeds of other plants which rely on light structures maximizing air friction to retard the descending motion.

Structure and geometry. The impatiens fruits can be divided into two groups depending on the geometry of the fruit: fusiform or linear. Fusiform fruits are generally to be found in tropical areas. The pods are shorter and rounder than the linear fruit. Linear fruits are native to temperate climates. The *I. glandulifera*'s shell-like seedpods belong to this latter category. They are 1.5–3.5 cm long, divided into five identical elastic valves containing 4–16 seeds (Chapman and Gray 2012; CABI 2016).

The valves are connected to each other by a thin tissue, which tears up during an action called 'dehiscence', as the mature seedpod opens. Dehiscence used for seed dispersal is also called ballistochory (Del Campo and Whitaker 2008). The fracture occurs contemporarily along all five lines connecting the valves: it starts at the top of the fruit, detaching it from the plant, and proceeds downwards to the other extremity of the pod leaving the valves connected at the two tips (Del Campo and Whitaker 2008). The upper 3/4 of the valve have a rigid triangular cross-section,

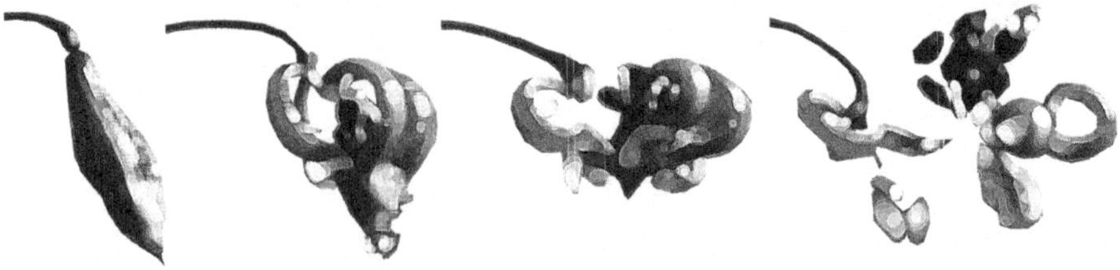

Fig. 3 Sequence of Himalayan balsam seedpods tearing up and throwing off their seeds

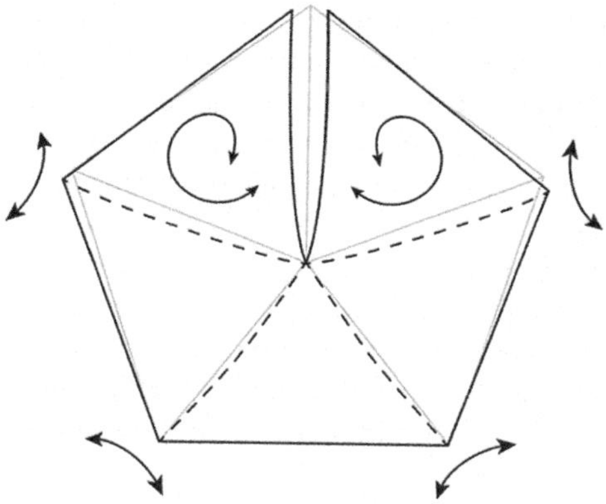

Fig. 4 Schematic view of the transversal cross-section of a pod, showing the pentaradial coupling between valves. As a first tissue cracks, equilibrium is lost and the valves curl cracking the other tissues as well (from Deegan 2012)

widening and thickening along the length, the lower 1/4 consists of a floppy membrane. The relaxed state of the rigid part of the valve is curled up, however, while the pod is still intact all valves are kept straightened out and connected to each other in a state of tension (Deegan 2012) (Fig. 4).

Motion sequence and mechanism. The explosive reaction of the Impatiens' seed dispersal is due to a bursting fracture movement occurring almost simultaneously along all dehiscence lines. The motion is irreversible and can be summed up in 3 different steps:

- Loading Phase—Elastic potential energy is built up during the pods' maturing process. The inner surface of each valve is kept way shorter than the outer side so tension is accumulated in the first one and compression is built up in the outer one as the closed pod holds the inner and the outer segment at the same length (Del Campo and Whitaker 2008). As the fruit ripens, tension maximizes.
- Triggering Phase—As the tensed pod is touched, one of the tissues between the valves cracks, balance drops leading to the contemporary cracking of all tissues (Deegan 2012) and to the simultaneous detachment from the plant.
- Relaxing Phase—The freed valves release their stored elastic energy instantly by separating from each other and coiling upwards. The elastic energy is converted into kinetic energy, and transferred to the seeds, which are effectively thrown away (CABI 2016). The maximum velocity recorded by Chapman and Gray (2012) was 12.4 m/s.

The curling of the valves is the key to the propulsion of the seeds. The coiling occurs inwardly and away from the axis of the pod, which allows the seeds to be forced upwards, transferring the momentum to the seeds (Del Campo and Whitaker 2008).

Releasing energy through fracture is a process, which needs to be finely calibrated not to dissipate the energy accumulated in the tearing process instead of transferring it into kinetic energy (Deegan 2012). The solution is the simultaneous cracking of all tissues, which minimizes energy loss: the seams open within 250 μs due to the geometry of the pod. The first crack drives a rotation of the other valves, cracking all other tissues. To assure the well functioning of this cascade of events, 80% of the tissue crack during the ripening process, long before the triggering occurs (Deegan 2012).

It is also interesting to notice that dehiscence detaches the pod from the plant, ripping it open from the top to the bottom. This on one hand ensures that the seeds are projected upwards instead of to the ground, greatly increasing the projectile distance. On the other hand, if not detached from the plant, part of the kinetic energy would get lost on moving the mass of the stem of the plant, reducing the energy transferred to the seeds.

1.3 Fern Spores, Catapulting Mechanism of the Sporangium

Leptosporangiate ferns reproduce dispersing spores of less than 50 μm, which are carried away by air currents over great distances. In order to detach them from the mother plant, these ferns have evolved a cavitation catapult system, able to fling the spores away with a fully reversible movement occurring at 10 m/s (Llorens et al. 2015) (Fig. 5).

Structure and geometry. The fern sporangium is a capsule of about 0.3 mm in diameter (Britannica 2016a) with thin walls, easy to tear up, containing the spores. An elastic ridge, fixed to the underside of the leaf, composed of a row of 12–25 differentially thickened cells called the annulus, runs on one side of the case. The annulus acts as a beam and is responsible for the spore ejection, working very much like a catapult.

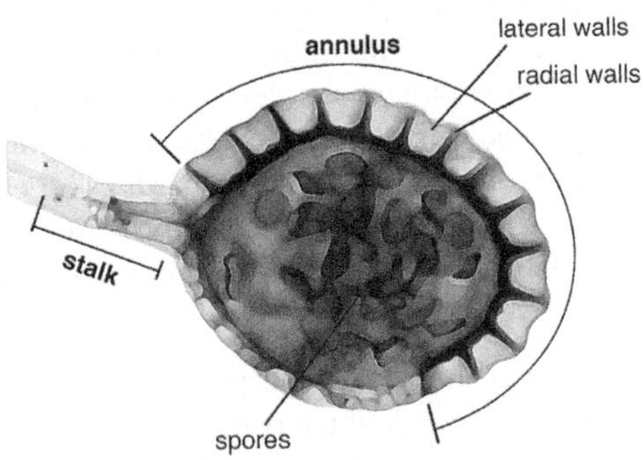

Fig. 5 Anatomy and parts of a closed fern sporangium

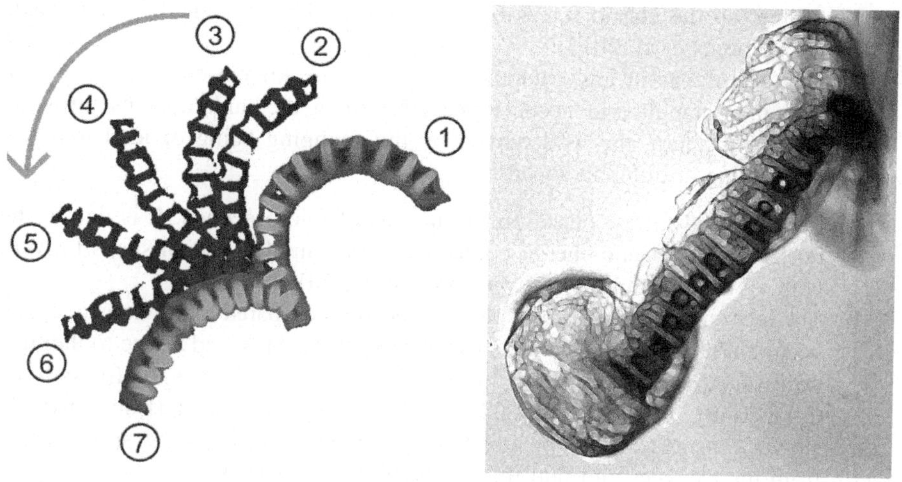

Fig. 6 Loading of the catapult in steps 1–7 (left) (modified from Noblin et al. 2012). Cavitation bubbles forming within the annulus, right before the release of the catapult (right) (modified from Llorens et al. 2015)

Motion sequence. The catapulting sequence can be described in three phases:

- Loading phase—elastic energy is stored in the annulus, which progressively bends backwards tearing up the spore capsule. As the deformation is progressive, the bending occurs relatively slowly opening up the spore case completely;
- Triggering phase—equilibrium is broken and the elastic energy gathered in the annulus is abruptly released (Britannica 2016a);
- Releasing phase—the annulus abruptly snaps back ejecting the spores (Noblin et al. 2012) (Fig. 6).

Motion mechanism. The mechanism has been largely known for over a hundred years and has a level of sophistication going beyond most man made technology. The mechanism responsible for the movement integrates motive force (water cohesion), energy storage (elastic annulus walls), triggering mechanism (cavitation) and breaking system (geometry) all in one element (Noblin et al. 2012).

- Loading phase—the water contained in the annulus cells evaporate creating negative pressure inside the cells (Britannica 2016a). The decrease in cell volume forces the thick radial walls to rotate towards each other, resulting in a bending movement of the ridge and the opening of the sporangium (Llorens et al. 2015). The negative pressure force bending the annulus counter works its material's elastic proprieties, putting it into tension exactly as a stretched rubber band.
- Triggering phase—when the water tension inside the cells is too large, cavitation occurs simultaneously inside the cells forming bubbles. When the bubbles exceed a certain size, the water tension is not enough to support the tension of

the spring, so the elastic forces of the annulus take over releasing the elastic energy (Llorens et al. 2015).

- Closing phase—as in any catapult, to transfer the kinetic energy to the projectiles, the catapult arm needs to be arrested in the middle of its throwing motion. In the fern, this is achieved by the swinging being divided into two phases closing at different speed:

  a. The first and fastest phase occurs in tens of microseconds. It is due to the release of the elastic energy contained in the annulus, which snaps forward with an acceleration of 105 g (Noblin et al. 2012).
  b. The movement in the second phase is slower, compared to previous one, and occurs in tens of milliseconds. The abrupt change in speed is due to the time necessary for water to filter through the porous annulus walls (poroelastic dissipation) (Llorens et al. 2015). This change in speed (from faster to slower) ensures an abrupt arrest, which effectively flings the spores away from the capsules and into the air (Noblin et al. 2012).

The mechanism is all the more interesting and rare because of its reversibility. Although this propriety does not seem to be exploited in nature, as the sporangium dies after ejecting the spores, immersing it into water for a few minutes can reverse the mechanism. Humidity allows the annulus cells to refill, making the cavitation bubbles progressively disappear and resetting the catapult to its original position (Llorens et al. 2015).

1.4 Horsetail's Spores, Walk and Jump for Spore Dispersal

Horsetails (*Equisteum*) are plants that typically thrive in highly humid habitats that periodically dry up under the effect of wind or sun. The spores are dispersed relying on the change in the surrounding relative humidity (RH) conditions, with two complemental modes.

Dispersal by walking is limited. The movement, which is driven by changes in geometry with every environmental humidity cycle, induces small displacements in a random direction, mostly allowing the spore to exit the sporangium, reorient and disentangle itself.

Dispersal through jumps is more efficient, as sudden thrusts propel the spores from the ground, raising the probabilities of a wind gust to catch on to the seeds.

Structure and geometry. The spores are small spheres of about 0.05 mm. Its coat divides, at maturity, into four ribbon-like appendages called elaters. The elaters exhibit hygroscopic proprieties and are believed to protect the spore by wrapping up around it in humid weather, and aiding dispersal by deploying in dry weather (Newcombe 1888):

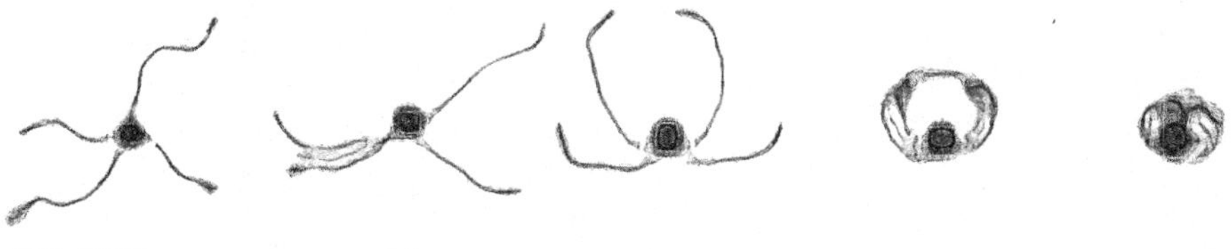

Fig. 7 Geometry of the elaters according to changing humidity conditions. The 'walking' is the combined effect of cycles of opening and curling of the elaters (modified from Marmottant et al. 2013)

- RH < 50% the elaters curl up as a screw;
- RH = 50% the elaters are straightened out, fully extended;
- RH < 75% the elaters start closing as the humidity rises;
- RH > 75% the elaters are wrapped around the spore (Fig. 7).

Motion sequence. During walks, the elaters open and close in response to repeated humidity cycles. As the Center of Mass oscillates in the various conformations, the spore tumbles around in random directions as the repeated opening and closing crates kinetic friction with the substrate. Repeated displacements achieve what are called the 'steps', which are small but important key displacements.

Jumps occur when the elaters suddenly dry up from a fully hydrated state, at speeds of 1 m/s and height of about a centimeter—which are very high, in comparison to the size of the spores.

A common spore dispersal sequence occurs as follows (Marmottant et al. 2013):

1. With recurrent variations in humidity, the spores achieve repeated movements, freeing themselves from the sporangium and eventually tangling up in clusters of spores,
2. With random steps, the spores reorient on the ground entangling more and more with other spores. With a repeated entanglement, the chances of the spores blocking the movements of each other rise, building up additional elastic energy in the clusters.
3. Spores with built-up elastic energy that are disturbed by external factors can jump off in random directions as they dry up or the limiting factors are removed (Fig. 8).

Motion mechanism. The mechanics are very similar to that of a laminate bicomposite material: two materials with different reaction coefficients are connected. As one of the layers reacts and the others doesn't, the material—which is usually produced in bands or stripes—bends.

Elaters have fully reversible hygromorphic capacities. The inner surface is more dense than the outer, with longitudinal microfibrils. The outer layer is instead highly porous and permeable to water, with a different expansion coefficient (Uehara and Kurita 1989). As the outer layer changes volume in the presence of rising humidity conditions and the inner layer remains unchanged, working against the expansion

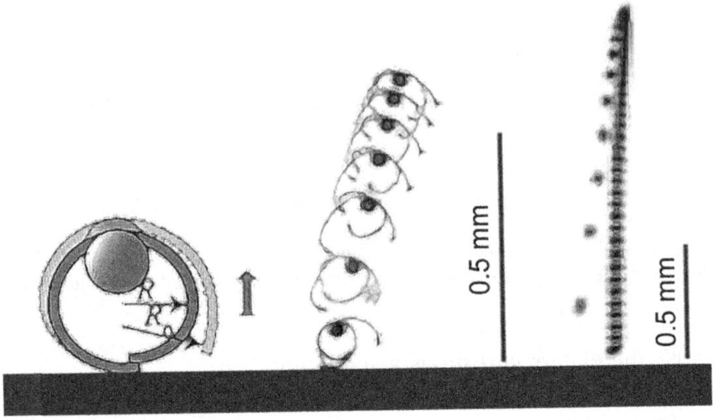

Fig. 8 Jumping mechanics of the elaters (modified from Marmottant et al. 2013)

and restrict the free movement of the outer layer, the stripes deform changing curvature. Any partial or complete restriction causes a corresponding force in the bending movement. The greater the expansion difference, the greater the effect (Ritter 2007).

1.5 Morning Glory, Photonastic Opening and Closing Hydraulics

Coordination of synchronized flower opening is often a quite complicated molecular mechanism, working in parallel with the normal growing processes of the plant. The timing of opening is precisely regulated by factors as light intensity, temperature and humidity for the nocturnal species. The triggering factor generally needs to act under certain duration before the flower opens, avoiding errors due to the interaction of other factors. Closing of the petals can be due to aging (giving permanent closure) or to an active process similar to the opening one (van Doorn and van Meeteren 2003). The Ipomoea, also called morning glory, is native to tropical and subtropical regions. The floral petal opening is quite fast, occurring within a range of minutes and is mainly due to a phytochrome reaction, responding to the changing light conditions making the flower opening at dawn (van Doorn and van Meeteren 2003) (Fig. 9).

Structure and geometry. The Ipomoea petals are helically furled and densely packed in a 10 cm long, 1–2 cm wide bud. The surface of the flower is made out of a single umbrella-like membrane wrapped 540° clockwise around the central axis in a logarithmic helix. Petal unfolding occurs within a few minutes in an outward spiraling opening, the final position of the petals depending considerably of their position in the bud (van Doorn and van Meeteren 2003). The unfolded flowers

Fig. 9 Three shapes in the Ipomoea flower (from left): closed bud, open corolla, toroid bud closure. In the fully opened flower, the midribs are visible in a lighter color under the flower membrane

bloom into a 8–15 cm radial pentameric umbrella (the lamina) with a tube-shaped base.

The size of the corolla is strongly influenced by the amount of turns in the helix, the more turns, the wider the flower. For this reason, flowers with constant spiral folding can potentially generate larger flowers, although the logarithmical helix packaging provides more stabile folded geometry (Schleicher et al. 2010) (Fig. 10).

A stiff band running along the petal's symmetrical axis, the midrib, characterizes the underside of the petals. In the bud state, the rolled up geometry exposes these band outwards, the ridges running adjacently to each other and effectively stiffening the bud and forming a shell-like protection. In the unfolded corolla, the band stretches the surface wide open and helps supporting the spanning structure by

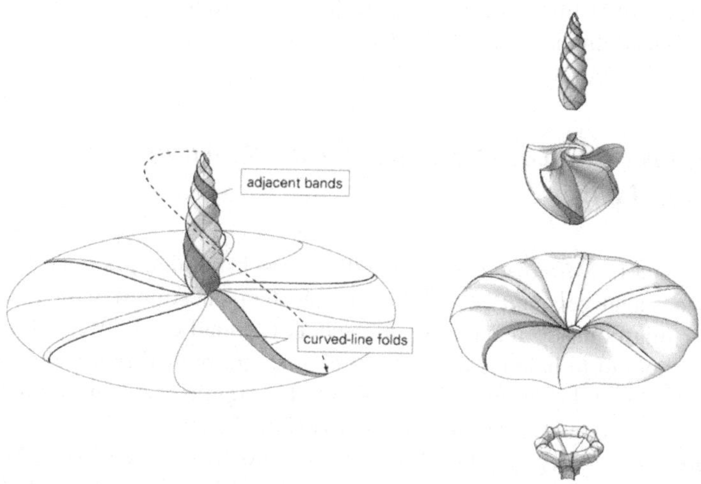

Fig. 10 Geometry and opening sequence of the Ipomoea (modified from Schleicher et al. 2010)

forming a large cantilever funnel stabilizing the surface through its double curvature rolled beam able to carry the load of the pollinators landing on it. Furthermore, each rib displays an S-shaped curvature along its length, providing much more stiffness than a linear beam (Schleicher et al. 2010).

Motion sequence. Two different opening motions can be identified in the Morning glory flower: a helical unrolling around the central axis, and an inwards curling of the petals. In the first case, the unfolding sequence of the densely packed bud depends on a contemporary helical unrolling of the structural bands and the unwrapping of the petal surface. Not being able to use the geometrical overlapping between petals to open the flower, the Ipomoea undertakes several complex structural changes folding, bending and expanding the corolla.

In the second case, the flower closes at sunset by curling the surface inwards by more than one complete turn. This vertical curling is fully reversible, as the mechanical tension in the petal drops as consequence of lower light conditions, the bud undertakes a toroidal shape (Schleicher et al. 2010).

Motion mechanism. Opening of flowers depend on four different kinds of petal movements which mostly occur in combination (van Doorn and van Meeteren 2003): reversible ion accumulation, localized cellular death (irreversible movement), water loss/water gain, differential cell-growth. Most characters occur especially in the combination with the latter kind of phenomenon, as the cells undergo a temporary elongation or swelling. In Ipomoea, the mechanism of flower opening and closure is more due to motion in the midrib than in the petal lamina. As a group of epidermal cells located all along the midrib undergo dynamic structural changes (as shape modification, enlargement and shape modification) they affect the internal turgor pressure driving the opening or the closing of the flower (Kaihara and Takimoto, 1981; Phillips and Kende 1980). Also the direction of the folding and unfolding is driven by the distribution and cell orientation of the tissues, which appears to be different in the midrib and in the lamina: fibers in the band follow the rib's S-shape, while fibers on the petal surface are arranged parallel to the rotation axis of the bud (Schleicher et al. 2010).

1.6 Mimosa Pudica, Defensive Nastic Reaction in the Leaves

Mimosa pudica L. (*Mimosaceae*), also called "touch-me-not", "shame" or "humble plant", is a sensitive plant responding to mechanical stimuli with a rapid seismonatic or thigmonastic reaction of its leaflets (Ahmad et al. 2012). Evolution of this trait is thought to balance the risk of being eating by herbivores (de Luccia and Friedman 2011), at the expense of a substantial energetic cost and interference with photosynthesis (Amador-Vargas et al. 2014) (Fig. 11).

Structure and geometry. The M. pudica is a prostrate or semi-erect bushy plant which grows up to 1 m with about 12–20 pairs of bi-pinnate palmated sensitive

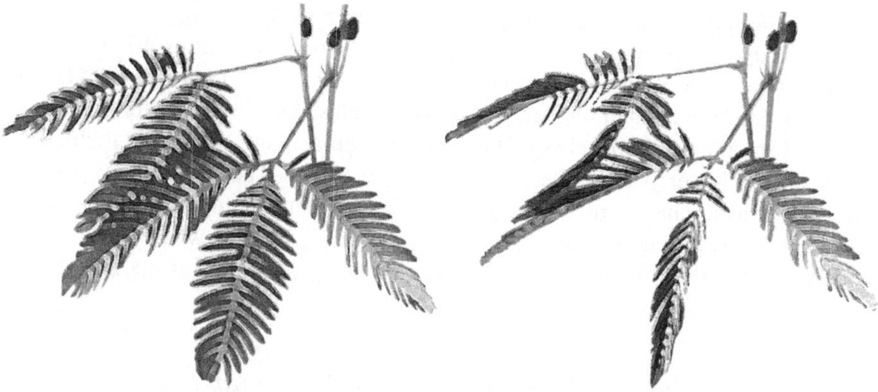

Fig. 11 A *Mimosa pudica* before (left) and after (right) triggering the thigmonastic reaction

leafs, each in turn composed of 10–20 pairs of 0.6–1.2-cm long 0.3–0.4-cm broad oblong glabrous leaflets with acute apex and rounded base, also called “petioles” (Ahmad et al. 2012).

Motion sequence. Movement can be initiated by many different kinds of interaction: touch, blowing, shaking, electrical chock, heat, rain, light, etc. Generally the movements observed are of two different kinds:

- a slow, nyctinastic, periodical movement controlled by the plant’s biological clock in response to daylight and night hours;
- a very rapid reaction in response to punctual stimulation or stress.

In the second case, action-reaction is immediate: the leaf rapidly folds completing the reaction in 1–2 s (Allen 1969). If no further stimulation occurs, the leaf opens up again after a few minutes (de Luccia and Friedman 2011). The more intense the stimulation, the longer the interval before opening (Ahmad et al. 2012) (Fig. 12).

Motion mechanism. The force-velocity range of the contraction mechanism has been compared to that of the typical animal muscle movements and have been

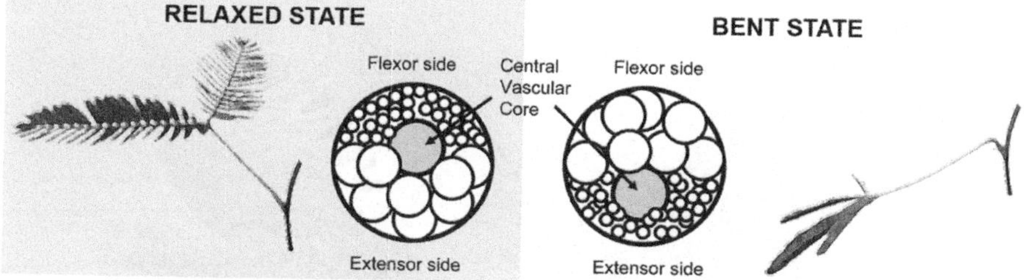

Fig. 12 Schematic diagram of the section of a pulvinus in the relaxed state (before stimulation), and during bending (after stimulation) (modified from Volkov et al. 2010)

found to be not only remarkably similar, but actually more efficient in the mechano-chemical energy conversion process (Balmer and Franks 1975).

The bending movement is carried out by the pulvinus—the joint-like thickening at the base of the leaflets—and not by the petioles themselves. When disturbed, specific regions of the plant's stem release chemicals, which activate so called "motor cells" located in the lower part of the pulvinus (Allen 1969). The initially swollen cells undergo a drastic change in volume, collapsing through a temporary loss of turgor pressure, with a consequent curvature of the pulvinus in the part where these cells are located, causing the folding of the leaf (Ahmad et al. 2012; de Luccia 2012).

1.7 Venus Flytrap, Thigmonastic Hunting Snap-Trap Mechanism

The Venus flytrap (*Dionanea muscipula*) is a carnivorous angiosperm, which has evolved its leaves as snap traps at the expense of part of its photosynthetic capacity (Hutchens and Luken 2009). The snapping motion is one of the most rapid movements in the plant kingdom, closing in about 0.3 s at a speed of 100 m/s (de Luccia 2012) (Fig. 13).

Structure and geometry. The plant is made of 5–7 leaves of a size between 3–7 cm (Pandolfi et al. 2014). Each leaf is made of two parts, the lower part—the lamina or footstalk—and a leaf like structure supporting the upper part—the trap. The trap is divided into two specular lobes held together by a blade. At the center of each lobe, next to a red pigment attracting insects, 3–5 hairs sensible to mechanical stimuli are arranged in a triangular pattern (Volkov et al. 2008). The edge of each blade is lined with hair-like protrusions (cilia), which interlock as the leaves closes, preventing all but the very smallest prey to escape (Yang et al. 2010).

Trapping sequence.

1. Open state, is a silent and stable phase with no observable movement. As an insect lands on the upper surface of the leaves and rubs against the trigger hairs, the trap enters an activation phase (Jaffe 1973). The hairs need to be stimulated

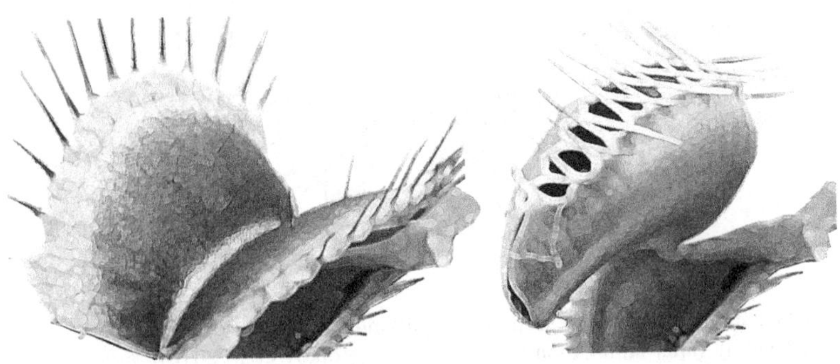

Fig. 13 A Dionanea muscipula in the open trap phase, and in the semi-closed trap phase

at least two times within 30 s to reach the threshold value for the trap to close. This prevents the plant to close when triggered by rain or wind, which are typically conditions below the stress range (Yang et al. 2010).

2. The trap snaps shut in a fraction of a second, or in fact into a semi-closed state: lightly ajar, in a state between opened and closed, for several minutes. This "decision-making" phase is adaptively advantageous, preventing the high energetic costs of secreting digestive enzymes into an empty trap (Yang et al. 2010). At this point, two different outcomes are possible. If nothing or non-nutritive material has been captured and no further stimulation of the trigger hairs occurs, the trap returns to the fully open state in 1–2 days. However if a prey has been captured and continues moving stimulating the trigger hairs, the trap proceeds to the fully closed state.
3. Closed state, with a new geometric equilibrium. The lobes seal tightly and remain closed for up to two weeks, during which the prey is digested (Volkov et al. 2008).
4. Reopening of the trap for the next victim. During the loading of the snapping mechanism, the leaves change shape in a very slow process—from flat, to concave and finally to convex (Volkov et al. 2008) (Fig. 14).

Trapping mechanism. Although this plant has been object of study and fascination for over two hundred years, the mechanism behind its fast nastic movement is still poorly understood. The most recent theory (the snap-buckling theory)

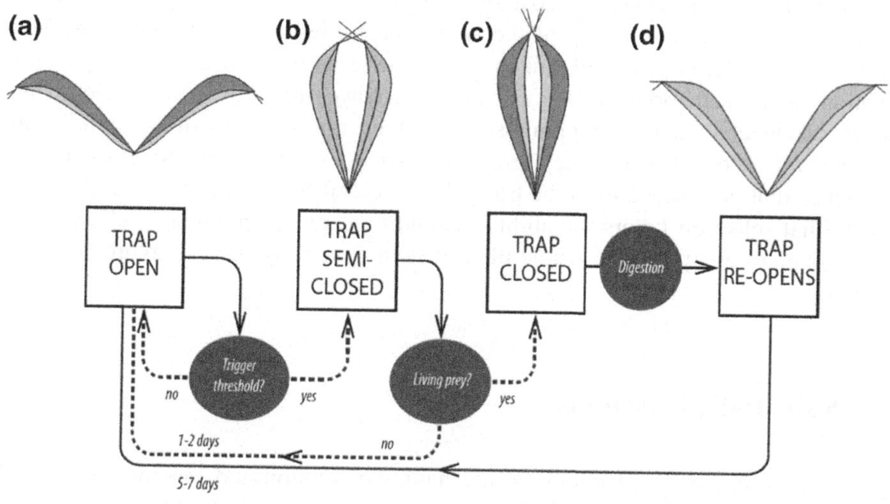

Fig. 14 Conceptual diagram of the snapping phases and the turgid layers of the leaves (based on Yang et al. 2010). In the first phase, the inner layer is thicker than the outer layer; in the second phase both layers have equal volume; in the third phase the outer layer is thicker than the inner layer; in the fourth and last phase the surfaces are minimal with zero mean curvature

suggests a combination of a geometric activation mechanism—a shape and curvature deformation resulting from a snap-buckling instability—and a chemical reaction—with a sudden change in cells' volume through a hydro elastic curvature mechanism (Pandolfi et al. 2014).

According to the snap-buckling theory, the driving force in the snapping is the elastic curvature energy stored in the curved pressurized leaves, loaded like a spring with a pressure difference between the upper and the lower layers of the leaves in a stable state. The stimulation of the trigger hairs opens up pores in both hydraulic layers, stability drops drastically and, in order to regain equilibrium, the convex geometry of the leaf snaps closed into its opposite concave (closed) position. The movement occurs though changes in curvature rather than movement of the leaves (Volkov et al. 2008).

These kind of reactions based on hydraulics and mechanics represents an alternative to muscle-activated kinetics. The differential turgor pressure is actively regulated by the plants and allows fast thigmonastic movements after a longer period of building up the pressure (Volkov et al. 2008).

Scaling. Out of about 600 existing carnivorous plant species, the *D. muscipula* is among the very few together with the *Aldrovanda vesiculosa* L., which has a nastic trapping mechanism. In fact, the high cost of evolving kinetic response must be balanced with the trap size, efficiency, prey availability, prey nutrient content, and season, among other factors (Hutchens and Luken 2009).

The relationship between prey and trap size has long been thought to favor the capture of larger preys, because of the space between the leaves' cilia: according to Darwin (1888), the energetic cost of capturing small insects would exceed the benefits, hence the possibility of the small and useless to escape the traps. More recent studies (Hutchens and Luken 2009), have however not found any confirmation of these theories. In the *Dionaea*, all leaves regardless of developmental stage produce traps and trapping success does not seem to differ significantly among trap sizes, although prey length is related to trap length. Also, there is no evidence that prey size affects the triggering effect of the trap. In the end, it is more likely that selection favors the ability of holding captured preys over allowing the small ones to escape as the whole plant benefits over any insect capture across all leaves.

2 Soft-Bodied Systems

In this part, examples of hydraulic mechanics of motion and locomotion through compartment-characterized soft-bodied structures in animals are shown.

Soft-bodied structures are used both as body-structures by the organisms as a whole, built up on hydraulic-cushion systems and moving accordingly, and as organs or body parts with hydraulic characters such as feet, tongues, trunks, etc., which are used in soft-bodied and rigid structures alike. These structures range from single, multiple and to juxtaposed compartment systems.

Soft-bodied locomotion is a very diffused way for moving in insects. Not being controlled by rigid parts and defined folding points (joints), these animals are able to move in any body plane making the movements potentially very complex (Mezoff et al. 2004).

The case-study sheets explore the following soft-bodied systems, with a specific focus on one of its kinetic features:

Name of the organism	Kinetic aspect studied
Earthworm	Peristaltic locomotion
Caterpillar	Crawling and Inching locomotion
Asteroid tube feet	Hydrostatic locomotion
Butterflies and moths' proboscis	Coiling/uncoiling dynamics

2.1 *Earthworm, Peristaltic Locomotion in the Lumbricus Terrestris L*

As most soft-bodied animals with hydrostatic skeletons, earthworms locomote by deforming their body segments.

Structure and geometry. The L. terrestris is substantially a cylinder-formed organism enclosing a fixed volume of liquid (Chapman 1950) with a musculature arranged circumferentially and longitudinally to control the diameter and the length of the cylinder by using the incompressible proprieties of the liquid to transmit motion and to provide antagonist force to re-extend the muscles (Kier 2012).

The tissue walls enclosing the liquid are divided into about 145 segments (septa) similar to each other, lacking localized specialization. The subdivision into septa allows localized control and variable movement patterns through the regulation of the flow of the liquid in the segments, as a sluice gate system. The body changes in length by changing diameter in the septa: the body walls can extend by more than 10% through contraction of the circular muscles (Kier 2012; Quillin 1999) (Fig. 15).

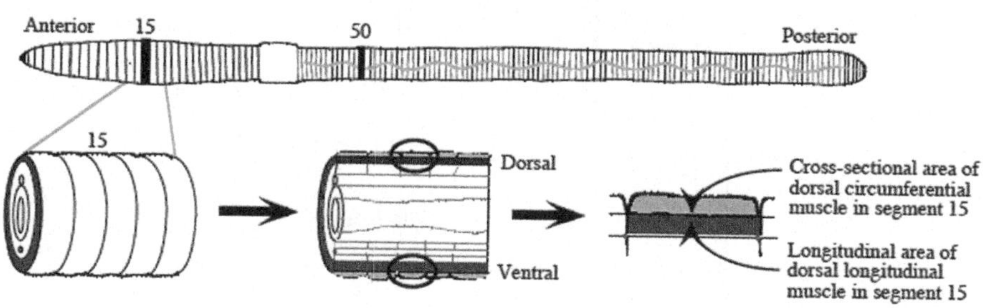

Fig. 15 Earthworm anatomy: longitudinal section of a *L. terrestris* with zoom-in on the dorsal circumferential and longitudinal muscles (from Quillin 2000)

Motion sequence. These soft-bodied crawlers typically use peristalsis for locomotion. Propulsion is achieved by letting rhythmic retrograde waves (waves progressing in the opposite direction to the locomotion) of muscle contraction pass through the whole body, starting from the anterior part of the animal. As a wave reaches the half of the body length, a new wave is initiated at the anterior end and the cycle is repeated over by each segment (Seymour 1969).

The expanded segments anchor the body to the substratum while the contracted ones advance, progressing with about 2–3 cm in each cycle. Speed equals the length of each stride multiplied by the frequency of the waves passing over the body (Quillin 1999; Gray and Lissmann 1938).

Motion mechanism. Earthworms use their internal body liquid to amplify the force, velocity and kinetic effect of their muscle-contraction by regulating the pressure in their body segments (Kier 2012). As the circumferential muscles of a segment contract while the amount of fluid it contains remains constant, the cylinder is elongated from thick to thin. As the longitudinal muscles contract, the shape

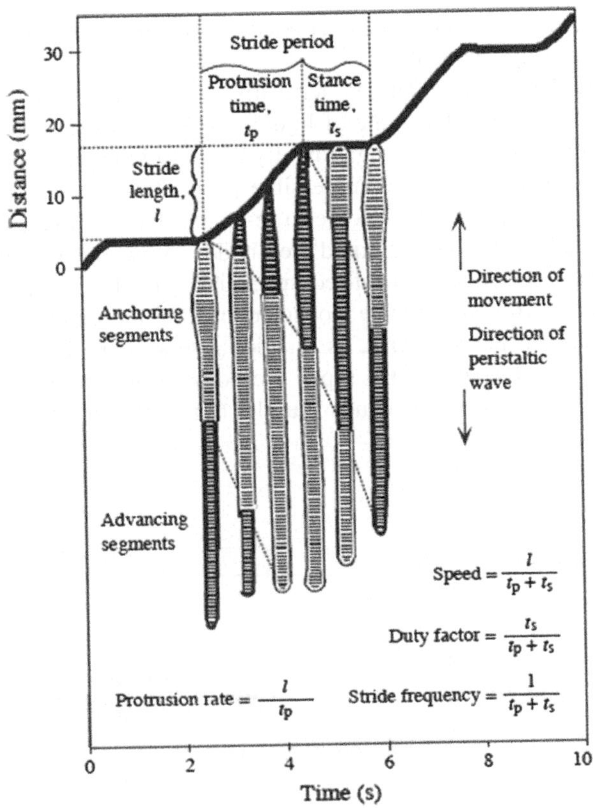

Fig. 16 Variables describing peristaltic locomotion (form Quillin 1999)

reverses to a thick and short cylinder. The pulsatile shifting between these two shapes in the septa result in a forward linear thrust of the body (Quillin 1999).

Longitudinal and circumferential muscles act as agonists: because each septa encloses a constant volume of liquid, contraction of one set of muscles result in the extension of the opposite set of muscles, force being transmitted through the fluid (Fig. 16).

Locomotory characters and gait. The stride length for a hydrostat would be the horizontal distance traveled during one cycle of peristalsis. Speed can be increased by taking longer and/or more strides per unit of time. However, as speed depends on the geometrical dimensions of the body, all earthworms keep similar speeds in relation to their size (Quillin 1999). Peristaltic locomotors can change speed but do not change gait. Although speed is generally a factor of high selective pressure both for predators and for preys, in the case of crawling, this locomotory technique do not seem very fit for fast movements as organisms generally choose non-peristaltic mechanisms for fast movements (Quillin 1999).

2.2 Crawling and Inching Locomotion in the Manduca Sexta and Other Caterpillars

Not all soft-bodied animals are hydrostats using incompressible fluids. Caterpillars have an extensive internal trachea containing, exchanging and compressing air. As the cavities are permeable and air leakages could compromise the efficiency of the force transmission, instead of losing efficiency through deformation, the caterpillars use the stiffness of the substrate to transmit forces (Lin et al. 2011).

Although being reliable, this locomotory technique does not allow fast movements. As speed has been of no evolutionary advantage for these animals (not being able to outpace their predators), other defensive mechanisms have evolved instead: aposematismic signals (warning signals as coloration, odours, sounds mimicking other species) (Casey 1991).

Structure and geometry. Caterpillars have three pairs of different legs: thoracic prolegs at the front end with a strong gripping-hook on each foot (clasper), abdominal prolegs more or less in the second half of the body-length, and anal prolegs in the rear (Mezoff et al. 2004). The body is mainly composed of one cavity without any septa or body segments. This allows body fluids and tissues to be moved along the longitudinal axis of the animal as the perimetral body segments stretch or compress, moving the caterpillar's Centre of Mass forward (van Griethuijsen and Trimmer 2009).

Motion sequence. Caterpillar locomotion (on horizontal or vertical plane) can mainly be dived into inching and crawling. Different species of caterpillars adapt one or both techniques, supposedly in relation to body-mechanics/size: large caterpillars keep their body close to the ground while small caterpillars can achieve self-supporting stiffness. In fact, some caterpillars (not the *Manduca sexta*) have shown to switch gait during growth (Lin et al. 2011).

Inching involves that the worm uses only the feet on its rear end. It casts the front part of its body to find an appropriate grip, then bends in the middle of the body-length and moves the posterior part to grip next to the anterior part. The movement requires an efficient structural control, precision and coordination (Lin et al. 2011).

Crawling locomotion is achieved with a progressive series of anterior-posterior body-wave contractions (Casey 1991). One cycle is characterized by three main steps (van Griethuijsen and Trimmer 2009; Casey 1991):

- the tip of the abdomen is moved forward and the terminal prolegs anchor the body to the substrate;
- an anteriograde wave of muscle contraction proceeds from the rear end;
- as the wave reaches the anterior part of the animal, the legs unhook and anchor the body further away (Fig. 17).

As in vertebrates, to change speed, caterpillars modify the cycle frequency by increasing stride length, stride frequency or muscle force with a consequent increase of energy cost per stride. The typical movement patterns achieved and the

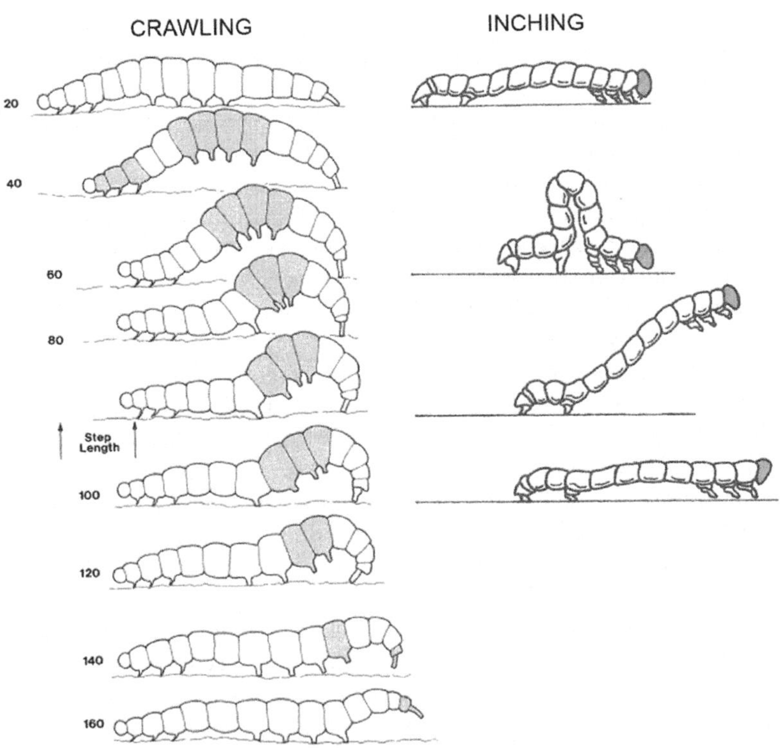

Fig. 17 Crawling locomotion in a caterpillar (modified from Brackenbury 1999), and inching locomotive sequence. In both figures the animals move from left to right

body morphology, constraining the animals to take short stride lengths (about 30% of that of animals with solid skeletons) that allow only locomotion at slow speeds (Casey 1991). Faster maneuvers can however be performed by reversing the waves in several gaits, which however brings potential instability (Brackenbury 1999):

- reverse walking, with inverted kinematics of the forward walking;
- reverse galloping, where all legs but the hooks are off the ground in each step;
- recoil and roll—a unique form of locomotion—where the caterpillar quickly bends inwards, the front touching the back, and rolls down from its support making use of gravitational potential force.

Motion mechanism. Caterpillars, as a hydrostatic system, separates from the standard constant body-volume constraints of most hydrostats for a number of reasons:

- Longitudinal extension is due to inter-segmental folds, not to stretching;
- Body pressure is variable during motion due to a fluid flow;
- Air can escape the volume through the trachea;
- The animal is not compartmentalized and cannot control pressure locally;
- The animal has leg-contact with the substrate and do not use the typical body-contact frictional-crawling type of Body Mass transfer;
- Mechanical behavior in the tissues is non-linear and anisotropic.

Caterpillars control their body stiffness mainly with pressurization of the body shell and muscle activation, employed simultaneously (Lin et al. 2011).

When crawling, contact with the substratum is mainly provided through hooking the claspers to the surface. As the wave of segmental contraction moves forward, muscles are activated to lift and unhook the prolegs so they can be moved forwards and hook to the next spot (van Griethuijsen and Trimmer 2009).

2.3 *Tube Feet, Hydrostatic Locomotion in Echinoderms*

Echinoderms have a hydrostatic skeleton and move through a multitude of small hydraulic feet, which can change shape and orientation by using the displacement of a fluid inside their cavity (Pritts and Rahn 2004). The feet are not only used for locomotion, but also for burrowing, feeding, breathing and sensing (Columbia Encyclopedia 2016).

A direct biomimetic application of these structures are pneumatic artificial muscles, which made of a flexible tubular membrane reinforced with fibers. Change of pressure in the tube causes it to extend or contract (Kier 2012) (Fig. 18).

Structure and geometry. All living echinoderms are pentaradially symmetric with an internal skeleton and a body cavity (coelom) with water-filled canals (ambulacra or water-vascular). In the sea stars in particular, the ambulacra is connected with the tube feet: hundreds of small cylinders or tentacle-like protrusions of the body wall extending in rows through holes in the skeleton (Britannica 2016b). On the tip of the tube feet are suction cups, which allow the starfishes to

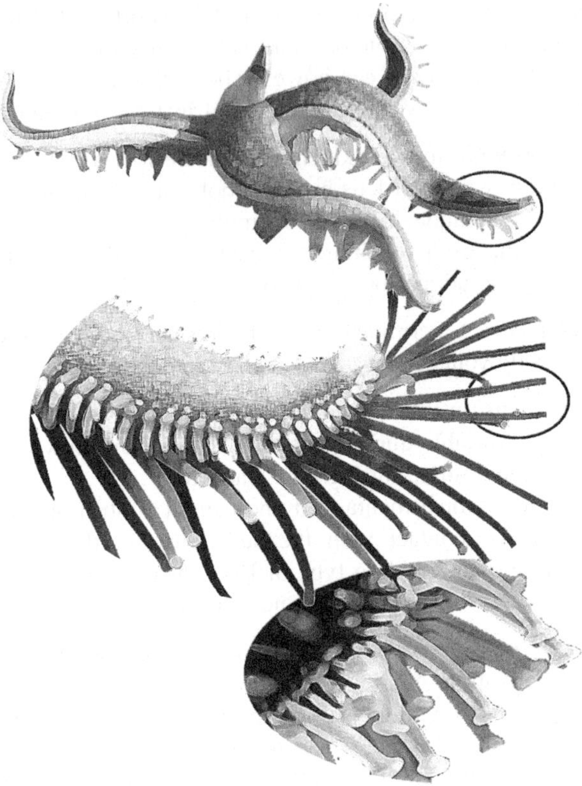

Fig. 18 Close-up of a walking asteroid and the tube feet

stick to a surface. Each cylindrical tube is connected to a muscular bulb (ampulla) aligned with the muscular bulbs of the other tube feet on the inner side of the ambulacra. The tube foot-ampulla system is connected and supplied with water through the radial canal through a one-way valve (Columbia Encyclopedia 2016; Pritts and Rahn 2004) (Fig. 19).

Most important for the motion and directing of the tube feet is the arrangement of circular and longitudinal muscle fibers able to contract selectively, antagonizing each other. The walls of the tube feet are reinforced with cross-fiber helical connective tissues, allowing shape change through extension and bending but preventing torsion and circumferential inflation (McCurley and Kier 1995; Kier 2012).

Motion sequence. Sea stars crawl around on rocky and muddy bottoms moving forward with any side of the body and reverse direction without having to turn around. Only one arm however becomes dominant in the locomotion, while the tube feet on all other arms move in the same plane as the lead arm (Columbia Encyclopedia 2016).

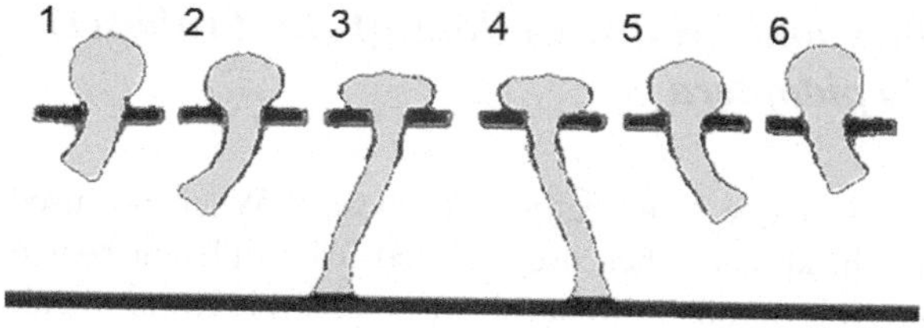

Fig. 19 Extension cycle of the tube feet

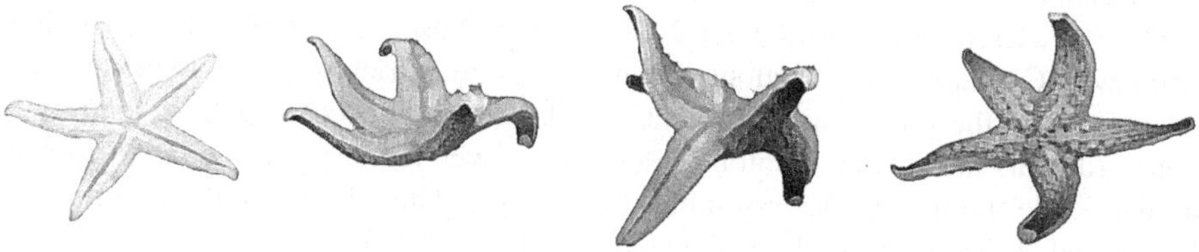

Fig. 20 Righting response sequence in a sea star

The tube feet are used as levers to move forward, or to stick to any other mechanisms that can pull it forward. They appear to move quite uncoordinated apart from small areas (Britannica 2016c).

Studies made on echinoderms do not reveal if any variations in gait can be taken with this type of locomotion, but other special movements have been recorded.

- Righting response. As in most echinoderms, the starfish's normal position is with the mouthparts facing a surface, if turned upside down, it performs a slow somersault using its arms and tube feet to return to its normal position;
- Burrowing. Several asteroid species bury themselves in sand or mud (Fig. 20).

Motion mechanism. The mechanism relies on the interaction of two connected pouches—the ampullae and the tube foot—which antagonize each other by alternatively inflating and deflating transmitting force through the incompressibility of the vascular fluid (seawater). Decrease in volume in the ampullae (through contraction) forces the fluid into the tube foot, resulting in its extension (McCurley and Kier 1995). As the muscles of the tube feet contract, the water is forced back into the ampulla which refills with the same liquid. The muscles in the wall of the tube foot controls its direction and bending movements. The importance of the helical arrays is mainly due to the fibers' angle of inclination, which transmits deformation to the tube it surrounds (Kier 2012).

2.4 *Butterflies and Moths, Coiling of the Proboscis in the Lepidoptera*

Lepidoptera have a unique trunk called a proboscis. When not used, it is rolled up (coiled) under the head, and when used to feed from different sources from liquid to sticky consistency, it is unrolled (Monaenkova et al. 2011; Kornev et al. 2009) (Fig. 21).

Structure and geometry. The Lepidopteran proboscis ranges from 0.1–30 cm in length and changes slightly in its structure depending on the feeding habits of the animal. The cross-section has an approximately elliptical shape, which continuously changes along its length, while the central food canal remains constant (Monaenkova et al. 2011; Krenn and Mühlberger 2002). The proboscis is composed of two linear strips with C-shaped cross sections (the galeae), symmetrically rolled up next to each other forming the food canal. Thin cuticles (legulae) connect the galeae, joining the upper (dorsal) and lower (ventral) sides. The lower legulae are similar to toothed hooks, while the upper ones resemble overlapping plates. Between the two runs the food canal (Lehnert et al. 2013; Eastham and Eassa 1955).

The proboscis can be extended only in the plane it is rolled up in: the sagittal. When extended, the dorsal surface of the proboscis bends into different convex shapes at about 1/3 of its length (Eastham and Eassa 1955).

Motion sequence and mechanism. The motion performed is mainly of 4 types (Krenn 1990):

- Resting position. The proboscis is rolled up tightly against the ventral side of the head, the legulae interlocking between the coils.
- Unrolling occurs in 3–5 steps simultaneously on the whole length of the proboscis, as a liquid (hemolymph) is pumped into the galeae until they are fully extended.
- Steering movements are performed though the action of muscles variously arranged in the galeal core, allowing the precise steering of the tubes.
- Rolling up. Coiling of the proboscis starts, unlike the uncoiling, from the tip and progresses up to the base, probably because of the necessity to precisely and compactly package the organ. When rolled up, the spiral widens up until it touches the ventral part of the head and the cuticles interlock in the resting position (Fig. 22).

Fig. 21 Butterfly with a coiled proboscis (left); geometry of the coiled proboscis (middle); butterfly with an uncoiled proboscis sucking nectar from a flower (right)

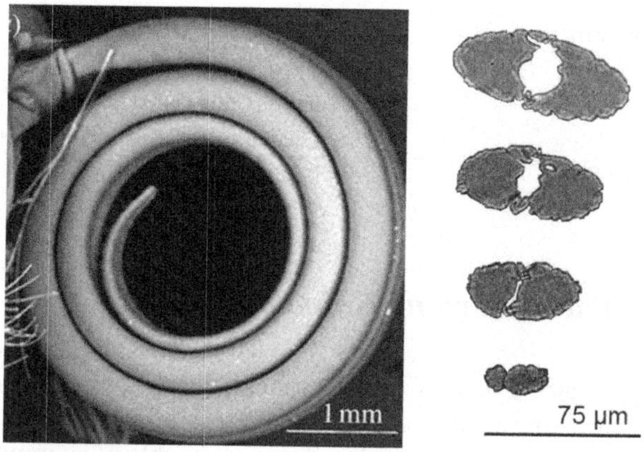

Fig. 22 Microscopy picture of a proboscis in the coiled resting state, and four cross-sections of a proboscis from the base to the tip (top-down) (modified from Monaenkova et al. 2011)

Calculations have shown that although the proboscis looks very much like a drinking straw, the animal would need to produce a suction pressure of over 1 atm to successfully suck up viscous sweet liquids. Other insects use a system of flexing pulsation of the proboscis to aid the pressure (Tsai et al. 2014; Knight 2014; Kornev et al. 2009).

3 Rigid Systems

This part shows few examples of animals using mechanics of rigid systems, vertebrates and shell-animals (endo- and exoskeletons). The four case studies have consciously been chosen because of the relative simplicity and clarity of the specifically studied aspects, to assure an easier potential biomimetic technology transfer of the analyzed concepts.

These systems all rely on displacement through the actuation of different variations of the lever system: the bivalves use a simple lever system, opening and closing the shell; the Insect wings have evolved an advanced lever system to fly at high frequencies; the click beetle uses an explosive thrust mechanism; and the snake makes the most out of the kinetic potential in the chain system using undulating locomotion.

The case-study sheets explore the following rigid-body structures, focusing on one of the reactive kinetic structures and mechanisms:

Name of the organism	Kinetic aspect studied
Scallop	Valve clapping locomotion
Insect	Indirect flight
Click beetle	Catapulting mechanism
Snake	Locomotive patterns

3.1 Scallop Valve-Clapping Swimming

Bivalves are freshwater mollusks (among others, clams, oysters, cockles, mussels, scallops, etc.), consisting of two symmetrical shells protecting the animal. Out of 17.000 known species of bivalves, only a few are capable of swimming: scallops in particular have evolved this capacity to escape predation (Tremblay et al. 2012) (Fig. 23).

Motion is achieved though a very basic lever system which relies on a streamlined light shell (the beam), a strong adductor muscle (the effort), and a very elastic hinge connecting the two shells (the fulcrum). By clapping the valves together, a water jet is expelled from the dorsal edge of the shell, providing thrust in the opposite direction to the opening of the valves (Denny and Miller 2006).

Structure and geometry. Scallops exhibit a very wide range of shell morphologies, from very streamlined to less hydrodynamic ones (Tremblay et al. 2012). All however rely on the same locomotion characters and body parts. The two rigid specular shells are connected though a hinge enclosing and protecting the soft-bodied mollusk, which actuates its exoskeleton through the action of one adductor muscle alone, providing contraction (Dickinson et al. 2000). A very important body part for the well functioning of the mechanism is the hinge

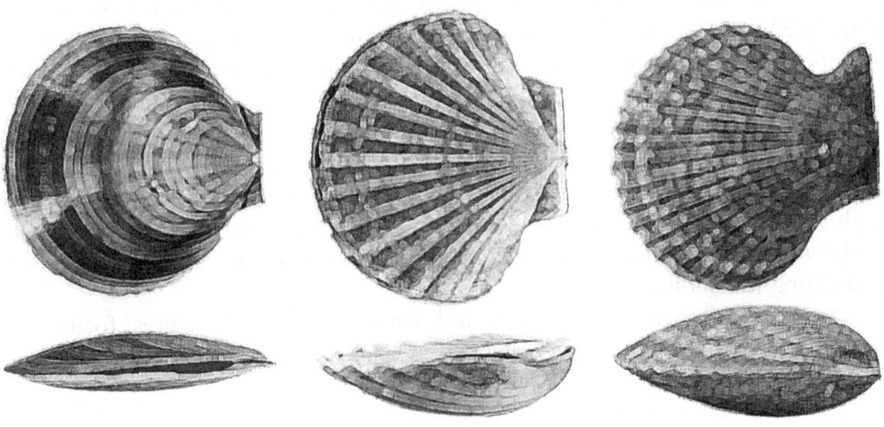

Fig. 23 Tree different types of scallop shells (top and side view). From the left: *Amusium balloti*, *Pecten Fumatus*, and *Mimachlamys asperrima* (modified from Tremblay et al. 2012)

ligament, which connects the two shell halves. This ligament has excellent elastic proprieties, acting as a flexible hinge to allow the smooth and repeated closure of the valves, and provides antagonist force to the only muscle the scallop possesses. While the outer part of the ligament holds the two valves together and in place on the hinge, the inner part is a highly elastic cushion, compressing as the shells are shut close, releasing elastic energy and extending the adductor muscle again as soon as it relaxes, opening the valves before the next contraction (Alexander 1966).

The bivalve's swimming ability is determined by factors as shell shape and the adductor muscle's metabolic capacity. Thrust is achieved through water jet propulsion; drag and lift depend on the swimming angle and speed, which are determined by the shell morphology. In scallops, good swimmers generally have light valves with smooth surfaces, and a high aspect ratio due to a slightly more convex upper valve, in comparison to the lower valve (Tremblay et al. 2012) (Fig. 24).

Motion sequence and mechanism. The motion sequence in scallop swimming can be summarized in three steps (Dickinson et al. 2000; Alexander 1966):

- Relaxed state. The shells are wide open, the adductor muscle and the spring ligament are relaxed;
- Activation. As the scallop is disturbed, it shortens the adductor muscle to snap the shells closed, compressing the elastic ligament and producing a water jet as a consequence of the fast shell closure, which thrusts the animal through the water;
- As the muscle is relaxed, the shell reopens through passive recoil of the elastic hinge ligament.

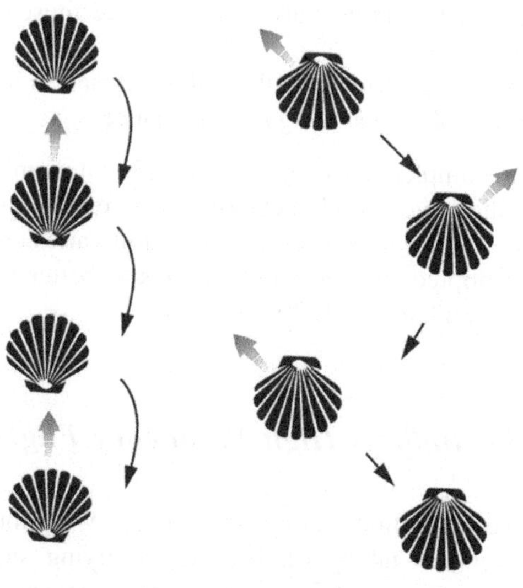

Fig. 24 Two modes of locomotion in the scallop. The black arrows indicate the direction of locomotion (proceeding top-down), the grey arrows the direction of the expelled water jet. The locomotive pattern on the left is an unidirectional jumping motion, where the scallop moves backwards. On the right, a zig-zagging locomotion as the scallop jumps forwards alternatively to the right and to the left side

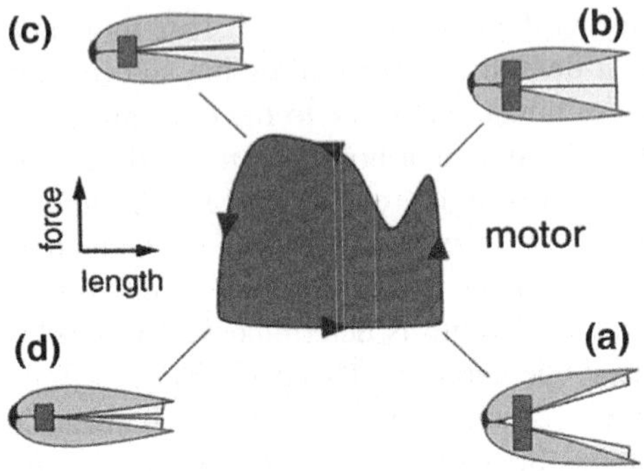

Fig. 25 Phases of muscle force generation in a scallop (modified from Dickinson et al. 2000). (**a**) The shell is at its maximal opening; (**a–c**) the muscle is shortened to drive a jet of water providing motion; (**d**) the shell is closed; (**d–a**) force drops, the elastic ligament reopens the shell and straightens the muscle

Swimming is achieved by repeating the opening/snapping motion several times per second, with phases of gliding through the water in between.

The adductor muscle in the scallop is particularly strong in proportion to its body mass, in comparison to not-swimming bivalves: the pressures created by the muscle can reach up to 2 kg/cm^2, producing the strong water jets on which the animals depend to locomote (Denny and Miller 2006; Trueman 1967) (Fig. 25).

The scallop performs two different types of contractions: phasic and tonic.

- Phasic contraction lead to rapid valve closure. The muscle is fast and highly energy demanding.
- Tonic contractions are used to keep the valves closed under a long time. The muscle is slower, but also low-energy maintenance.

These contractions complete each other: the tonic intervening when the phasic muscle is fatigued, allowing partial metabolic recovery (Tremblay et al. 2012). Timing and frequency of phasic and tonic contraction varies considerably between species, which have adopted specific combinations to better fit their morphology, habitat and lifestyle (Tremblay et al. 2012).

3.2 *Insect Flight, Indirect High-Frequency Flight*

Insects evolved active flight first among all animals. With multiple and complex aerodynamic maneuvers, as taking off backwards, flying sideways and landing upside down, they still remain unbeaten under many aspects (Young et al. 2009; Dickinson et al. 1999). Because of their small body size (20 μg–3 g), flying insects must beat their wings at a much higher frequencies than vertebrates to generate enough aerodynamic thrust (Nachtigall 2005).

Structure and geometry. The wings, thin membranes structurally straightened by a web of veins, are an outgrowth of the exoskeleton, enabling the insects to use a lever system to provide thrust. The wings are found on the second and third thoracic segment, and can occur in one or in two pairs (in that case described as fore- and hind-wings). Wings in two pairs are mechanically linked together and work in most cases as a single wing (Brodsky 1994).

With the lever system, a bigger wing area is activated through the action of the muscles on the one end of the structure. The rigid part of the wing (the beam), which mechanically transmits the muscle force to the wing surfaces, is connected to the thorax of the insect through an articulation (the fulcrum) and continues shortly inside the thorax (the effort), building a lever system similar to a seesaw. The flight muscles do not directly activate the wing beat as in vertebrates, but deform the thorax, applying the effort on the opposite end of the load, with the fulcrum in the middle (Nachtigall 2005).

Motion sequence. There are two main models of insect flight: leading edge vortex and fling-and-clap. Other gaits are hovering and gliding (Sane 2003; Fernandes et al. 1991).

Leading edge vortex is created through flapping the wings in two half-strokes:

- the down stroke starts up and backwards, moving forwards down;
- the wing is tilted around its axis, with the edge facing backwards, to minimize force on the upward stroke;
- the upstroke pushes the wing up and back;
- the wing is tilted again to the original position.

Fling-and-clap is used by very small insects. In this motion pattern the wings are clapped together above the insect's body and then flung apart. During the flinging motion, negative pressure is created above the animal, so air is sucked in creating a vortex over each wing creating lift. As the vortex moves over the wing, it also contributes to the clapping movement. The reason why bigger insects do not use this motion scheme, is because the repeated clapping would cause damage and wear to the wing (Dickinson et al. 1999).

In hovering, the insect stays on the same spot in the air, not moving forward. In this pattern, the wing beats twice as fast as in forward flight, requiring lift as well as stabilization (Sane 2003). As the insect starts flying forwards and flight speed increases, it also tilts its body and head down, minimizing the body area in contact with the airflow and the drag (Fig. 26).

Flies with two pairs of wings have evolved the two hind wings (halteres) to oscillate anti-phase to the front wings. These help the insects to identify rotations of the body during flight, and are therefore important for aerial maneuvers. The mechanisms underlying the synchronization between front and hind wings is not yet well understood, but seems to be due to specific mechanic linkages within the thorax of the insect (Deora et al. 2015).

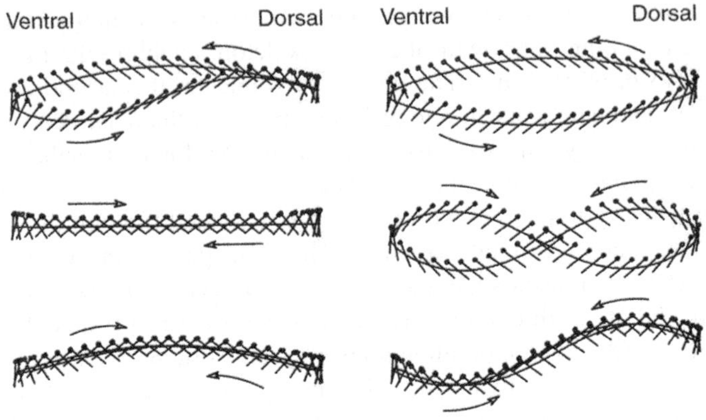

Fig. 26 Six possible variations of wing stroke patterns in insects (from Lehmann and Pick 2007)

Motion mechanism. The thorax works as a rigid box to which the wings are attached. As previously discussed, movement in the thorax brings movement in the wing thanks to two pairs of muscles:

- the first muscle set runs parallel to the dorso-ventral axis, connecting the thorax to the back of the insect. Contraction depresses the thorax area and raises the wing due to a reverse linkage between thorax and the wing;
- the other set of muscles run diagonally from the anterior of the thorax to the posterior floor. Contraction in the diagonal muscles elevates the thorax, extends the first set of muscles and provides the down stroke (Fig. 27).

In a few species, the down stoke is provided by an elastic recoil of the thorax as the muscles relax. Also, at the end of each down stroke, the wing needs to be slowed down to reverse for the upstroke. During down stroke, the muscles dissipate energy by converting it into heat, or (in a few species) converting it into potential energy for the upstroke through the action of an elastic pad (resilin). As the wing beats the upstroke, the resilin is stretched, giving back energy into the down stroke by snapping back (Biewener 2007).

Fig. 27 Actuation of the dorso-ventral muscles (left) and the tergosternal muscles (right), deforming the thorax and aiding the indirect motion of the wings

Because nervous systems cannot control stroke-by-stroke wing beats at such speeds, some insects (as Neoptera) have evolved a unique feature: the asynchronous flight muscle. This muscle is able to beat at higher frequencies (over 1.000 Hz) than the typical flight muscle (5–200 Hz) (Deora et al. 2015). A single nerve impulse causes the muscle fiber to contract multiple times, allowing the frequency produced to exceed by many times the rate at which a nervous system is able to cope with normal wing beat impulses (Sane 2003).

3.3 Click Beetle, Catapulting Mechanism in the Elateridae

Click beetles are small elongated hard shelled and (mostly) nocturnal insects named after the clicking sound which accompanies the bounding mechanism, that catapults them vertically by many body lengths. The clicking reaction is primarily used to turn the insect when it lands on the back, as its legs are too short to turn it around, but can also be used as defense against predators. The clicking can be repeated several times over, as the insect might not succeed to turn at the first thrust (Burton 2002; Ribak and Weihs 2011) (Fig. 28).

Structure and geometry. The bug's thorax is divided into two main parts: the prosternum (head) and the mesosternum (winged body part), a movable hinge connecting the two. On the central dorsal side of the prosternum the exoskeleton forms a spike-shaped protrusion, which slightly bends inwards resting on the opposite side of the mesosternum's exoskeleton. On the very edge of the place where the spike rests, the first section's exoskeleton forms a cavity. The hinge dividing the body into two subunits is locked by a membrane which opposes the action of a large muscle connecting the two halves and which is contracted to load the system with elastic energy (Evans 2010; Nachtigall 2005).

Motion sequence. As the beetle is lying on its back, it performs a violent nodding movement to jump vertically (over 30 cm in height and reaching an acceleration of about 3.800 m/s), in order to turn over in the air without using its legs (Evans 1972). Once the 'clicking' order has been given by the nervous system, follows a cascade of events (Ribak and Weihs 2011; Nachtigall 2005; Burton 2002):

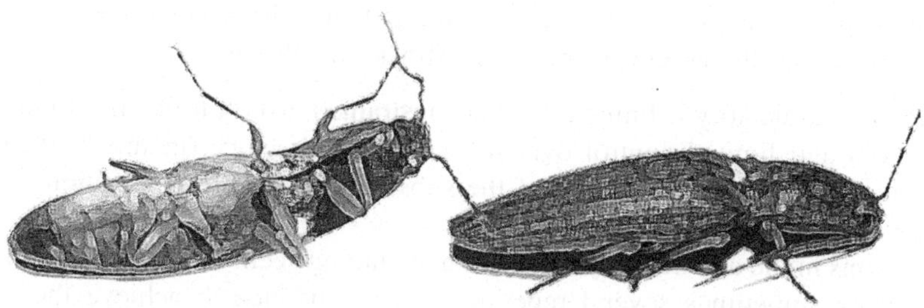

Fig. 28 To the left, a click beetle lying on its back, the spike on the prosternum visible as a dark line between the front legs. To the right, the beetle after the clicking and the jump, once rightened

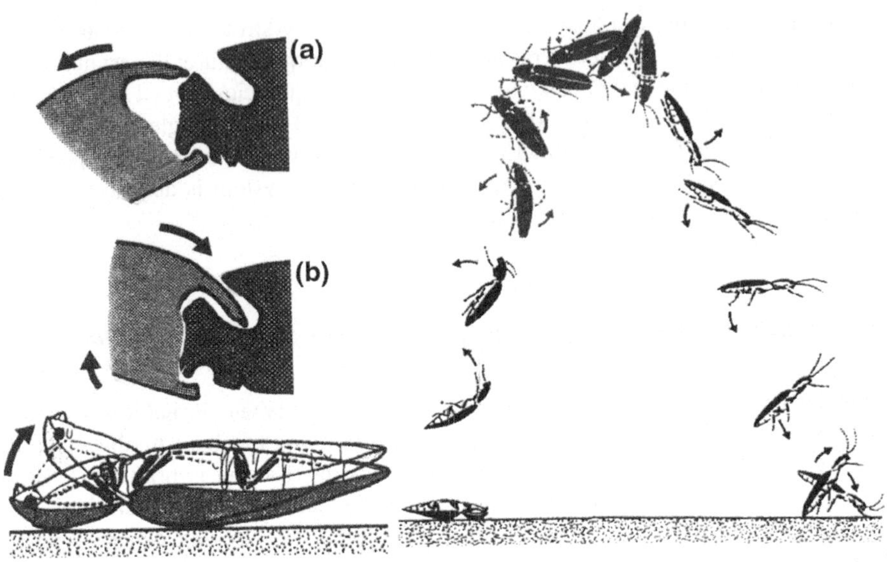

Fig. 29 Schematic detail of the clicking mechanism (modified from Nachtigall 2005). The spike is drawn up against the opposite surface and locked in that position (**a**). As the hinge is moved, the spike snaps back into the mesosternal pit (**b**), providing the kinetic thrust to fling the whole body into the air

- The ligament muscle contracts forcing the spike against he opposite surface, building up tension as the spike resists against the load. When the hinge mechanism is locked, the catapult is loaded with potential energy;
- The hinge is moved to allow the spike to slide over the edge of the surface. As the built up tension is released, the spike hits down violently on the bottom of the mesosternal pit producing the famous 'click' sound;
- The prosternum performs a fast backwards nodding movement, abruptly hitting the surface the beetle is lying on, flinging the insect vertically into the air (Fig. 29).

Many aspects regarding this mechanism are still a question mark to researchers, as they appear to be energetically unfit. The only parameter the beetles seem to control is the magnitude of the elastic force put into the system, influencing the takeoff speed and the height of the jump (Ribak and Weihs 2011).

1. The thrust trajectory is limited. With a constrained takeoff thrusting angle (79.9° ±1.56°) and limited control over the jumping trajectory (mainly vertical) the beetle must rely on the nature of the substratum to achieve variation in jump angles.
2. Rotations in the air and body orientation at landing seem to be poorly controlled, requiring sometimes several repetitions of the bounce to achieve the wished effect. While in the air, the beetle performs various rotations in order to get back on its feet, relying on a 50% random chance to landing on the right side.

3. More than 98% of the jumping force is directed vertically, to counteract gravity, without performing any substantial displacement. Slightly asymmetrical and tilted thrust angles would theoretically give better results both in the turning and in the escaping action.
4. The force put into the jump exceeds the movement necessary to perform the flipping motion. To turn back on its feet, a beetle would need to raise its body not more than one body length and perform a half revolution. Instead, the jumps exceed this minimum requisite by many times.

3.4 *Snake, Locomotive Patterns in Serpentes*

Although the body structure may seem very simple, snakes are extremely versatile in their environment. They are able to smoothly cross complex terrains as sand, stony and rocky grounds, woods and can even climb: a flexibility, which would be a challenge to any legged system (Biewener 2007).

Structure and geometry. Snakes' bodies can be imagined as the most basic forms of vertebrate skeleton, built with a central spine working like a roller chain, surrounded by organs and musculature following the longilinear shape of the flexible skeleton. Undulating animals typically exhibit segmented blocks of muscles running repeatedly next to each other from head to tail (Nachtigall 2005).

Snakes are known to adopt at least five main types of terrestrial locomotion. All snakes use most or all of these modes, depending on the terrain and the speed, and can even use a combination of two of these modes in different parts of the body (Moon 2010; Marvi et al. 2013).

- Lateral undulation is used on flat or structured surfaces;
- Sidewinding on sandy, smooth, slippery surfaces providing little support;
- Concertina in crevices wide enough for the snake to make two bends;
- Rectilinear locomotion for tight spaces where concertina cannot be used, for climbing on trees and moving silently to hunt;
- Slide-pushing for bursts of motion.

Motion sequence.

Lateral undulation (also called slithering or serpentine) is the most common and known mode of locomotion in snakes. Waves of lateral bending are passed through the body from head to tail. As the snake moves, the entire body follows the movement initiated from the head and neck, as the cars of a train. As the snake pushes its body alternatively to the sides, the lateral force vectors counteract each other's lateral component leaving a resultant vector that propels the snake forward at speeds around 10 cm/s (measured in a Boa by Marvi et al. 2013). As the snake's body finds an uneven terrain, as a rock or wood, the muscles deform locally grasping the object to control to the movement even better (Biewener 2007).

Sidewinding has a similar bending pattern as the lateral undulation, but provides a sequential static (instead of sliding) contact with the ground while the other parts of the body are lifted off the ground and down further away. The snake travels roughly diagonally relatively to the tracks formed on the ground, which are a very typical set of parallel straight lines (Moon 2010).

Concertina is an accordion-like locomotion where the body is pulled into two sets of bends with a part of the body straightened out in between. As the back bends grip to the ground, the front part of the body is extended. As the front part grips to the ground, the back part is pulled up into bends again (Marvi and Hu 2012; Biewener 2007; Day 1981).

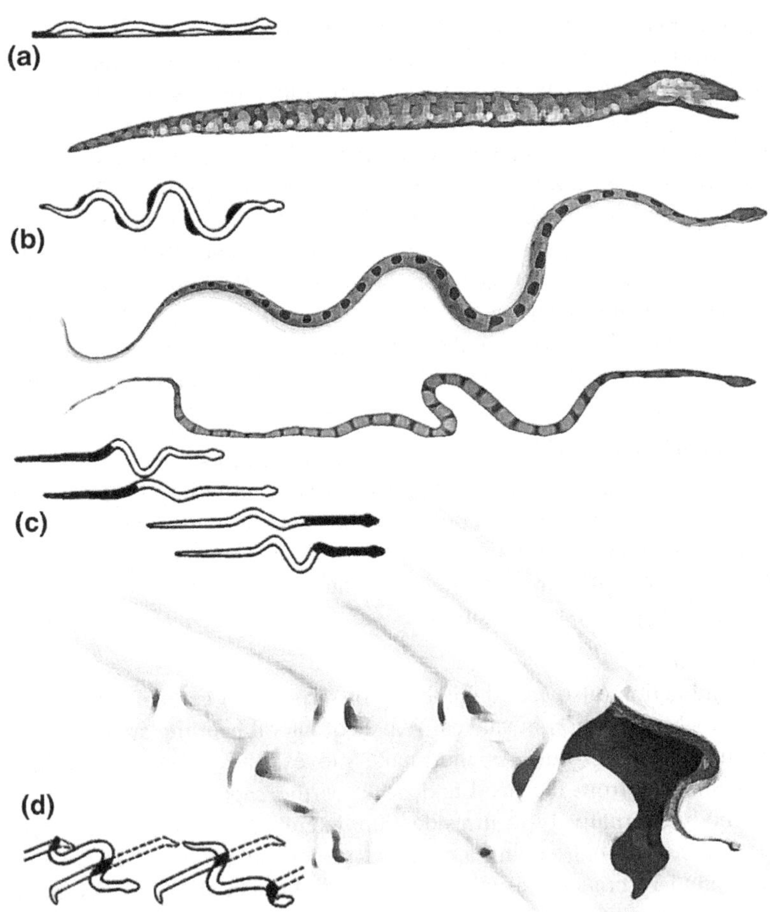

Fig. 30 Four types of snake locomotion. (**a**) rectilinear, (**b**) undulatory, (**c**) concertina and (**d**) sidewinding. In black on the schemes, the areas of contact with the ground providing support

Rectilinear locomotion is mostly used by large and heavy snakes to move slowly (0.2–6 cm/s in a Boa) and silently for hunting. It is one of the only or very few types of vertebrate locomotion not involving any direct action of any rigid part of the structure in the form of levers (Lissmann 1950): a part of the belly is alternatively lifted 1 mm from the ground, pulled forward, down, and backwards pulling the body forward. The cycle is repeated over simultaneously at many points along the body (Marvi et al. 2013).

Slide-pushing is used to escape quickly. Irregular bends of vigorous undulations push the ground at different points. Although the body almost certainly slips on the surface, the pushing force is strong enough to move the animal forward. The resulting progression over the ground is an irregularly slipping and sliding motion. Friction is the most important component of the motion although there might be occasional moments of static contact (Moon 2010; Guo and Mahadevan 2008) (Fig. 30).

Motion mechanism.

In lateral undulation, the dorsal muscles are alternatively activated on opposite sides of the body in each bend, from the convex part of a bend up to the next concave bend (Guo and Mahadevan 2008).

In sidewinding, muscles are activated in a similar way to lateral undulation with the addition of some muscles, which also act bilaterally to lift the trunk.

In concertina, blocks of muscles are activated simultaneously in the parts of the body providing support in the bendings.

Rectilinear locomotion requires a bilateral activity of the muscles connecting the skin in contact with the ground to the spine transmitting the movement. One set of muscles pulls the scales up and forward, another pulls them down and back (Lissmann 1950).

Muscle activity in the slide-pushing type of locomotion appear to be unknown at the moment (Moon 2010).

References

Ahmad H, Sehgal S, Mishra A, Gupta R (2012) *Mimosa pudica* L. (Laajvanti): an overview. Pharmacog Rev 6(12):115–124. https://doi.org/10.4103/0973-7847.99945

Alexander RMcN (1966) Rubber-like properties of the inner hinge-ligament of pectinidae. J Exp Biol 44(1):119–130

Allen RD (1969) Mechanism of the seismonastic reaction in mimosa pudica. Plant Physiol 44 (8):1101–1107. https://doi.org/10.1104/pp.44.8.1101

Amador-Vargas S, Dominguez M, Leon G, Maldonado B, Murillo J, Vides GL (2014) Leaf-folding response of a sensitive plant shows context-dependent behavioral plasticity. Plant Ecol 215(12):1445–1454. Springer, https://doi.org/10.1007/s11258-014-0401-4

ARKive (2016) Himalayan balsam (Impatiens glandulifera). Accessed 20 April 2016 from http://www.arkive.org/himalayan-balsam/impatiens-glandulifera/

Balmer RT, Franks JG (1975) Contractile characteristics of *Mimosa pudica* L. Plant Physiol 56 (4):464–467

Biewener AA (2007) Animal locomotion. Oxford University Press, Oxford, UK. ISBN 978-0-19-850022-3

Brackenbury J (1999) Fast locomotion in caterpillars. J Insect Phisiol 45(6):525-533. https://doi.org/10.1016/s0022-1910(98)00157-7

Britannica (2016a) Fern, fern spore. Accessed on 26 March 2016 from http://www.britannica.com/plant/fern/Ecology#toc49915

Britannica (2016b) Echinoderm. Accessed on 18 April 2016 from http://www.britannica.com/animal/echinoderm

Britannica (2016c) Locomotion. Accessed on 18 April 2016 from http://www.britannica.com/topic/locomotion/Bottom-locomotion

Brodsky AK (1994) The evolution of insect flight. Oxford University Press, Oxford, UK. ISBN 0-19-854681-5

Burton R (ed) (2002) International wildlife encyclopedia, 3rd ed., vol 4. Marshall Cavendish Corporation, pp 466–467

CABI (2016) Impatiens glandulifera (Himalayan balsam). In: Invasive species compendium. CAB International, Wallingford, UK. Accessed 20 April 2016 from http://www.cabi.org/isc/datasheet/28766

Casey TM (1991) Energetics of caterpillar locomotion: biomechanical constraints of a hydraulic skeleton. Science 252(5002):112–114. https://doi.org/10.1126/science.252.5002.112

Chapman DS, Gray A (2012) Complex interactions between the wind and ballistic seed dispersal in *Impatiens glandulifera* (Royle). J Ecol 100(4):874–883. https://doi.org/10.1111/j.1365-2745.2012.01977.x

Chapman G (1950) Of the movement of worms. J Exp Biol 27(1):29–39

Clements DR, Feenstra KR, Jones K, Staniforth R (2008) The biology of invasive plants in Canada 9. Impatiens glandulifera Royle. Can J Plant Sci 88(2):403–417. https://doi.org/10.4141/CJPS06040

Columbia Encyclopedia (2016) Echinodermata. Accessed 18 April 2016 from http://www.encyclopedia.com/reference/encyclopedias-almanacs-transcripts-and-maps/echinodermata

Darwin C (1888) Insectivorous plants, 2d edn. Revised by Francis Darwin, John Murray, London

Day MH (ed) (1981) Vertebrate locomotion, symposia of the Zoological Society of London no. 48. The Zoological Society of London, Academic Press Inc., Oxford, London

Deegan RD (2012) Finessing the fracture energy barrier in ballistic seed dispersal. Proc Natl Acad Sci USA (PNAS) 109(14):5166–5169. https://doi.org/10.1073/pnas.1119737109

Del Campo L, Whitaker DL (2008) The biomechanics of ballistochory in impatiens pallida. Pomona Senior Theses, Paper, p 26

De Luccia TPB (2012) *Mimosa pudica*, *Dionaea muscipula* and anesthetics. Landes Biosci Plant Signal Behav 7(9):1163–1167. https://doi.org/10.4161/psb.21000

De Luccia TPB, Friedman P (2011) Boolean function applied to *Mimosa pudica* movements. Landes Biosci Plant Signal Behav 6(9):1361–1364. https://doi.org/10.4161/psb.6.9.16445

Denny M, Miller L (2006) Jet propulsion in the cold: mechanics of swimming in the Antarctic scallop *Adamussium colbecki*. J Exp Biol 209:4503–4514. https://doi.org/10.1242/jeb.02538

Deora T, Singh AK, Sane SP (2015) Biomechanical basis of wing and haltere coordination in flies. Proc Natl Acad Sci USA (PNAS) 112(5):1481–1486. https://doi.org/10.1073/pnas.1412279112

Dickinson MH, Lehmann FO, Sane SP (1999) Wing rotation and the aerodynamic basis of insect flight. Science 284(5422):1954–1960. https://doi.org/10.1126/science.284.5422.1954

Dickinson MH, Farley CT, Full RJ, Koehl MAR, Kram R, Lehman S (2000) How animals move: an integrative view. Science 288:100–106. https://doi.org/10.1126/science.288.5463.100

Eastham LES, Eassa YEE (1955) The feeding mechanism of the butterfly *Pieris brassicae* L. Philos Trans Roy Soc B, Biol Sci 239(659):1–43. https://doi.org/10.1098/rstb.1955.0005

Evans MEG (1972) The jump of the click beetle (Coleo ptera: Elateridae)—a preliminary study. J Zool 167(3):319–336. https://doi.org/10.1111/j.1469-7998.1972.tb03115.x

Evans MEG (2010) The jump of the click beetle (Coleoptera, Elateridae)—energetics and mechanics. J Zool 169(2):181–194. https://doi.org/10.1111/j.1469-7998.1973.tb04553.x

Fernandes J, Bate M, Vijayraghavan K (1991) Development of the indirect flight muscles of *Drosophila*. Development 113(1):67–77

Gray J, Lissmann HW (1938) Studies in animal locomotion, VII. Locomotory reflexes in the earthworm. J Exp Biol 15(4):506–517

Guo ZV, Mahadevan L (2008) Limbless undulatory propulsion on land. Proc Natl Acad Sci USA (PNAS) 105(9):3179–3184. https://doi.org/10.1073/pnas.0705442105

Hsiao TC, O'Toole JC, Yambao EB, Turner NC (1984) Influence of osmotic adjustment on leaf rolling and tissue death in rice (*Oryza sativa* L.). Plant Physiol 75:338–341. https://doi.org/10.1104/pp.75.2.338

Hutchens JJ, Luken JO (2009) Prey capture in the venus flytrap: collection or selection? Botany 87:1007–1010. https://doi.org/10.1139/B09-064

Jaffe MJ (1973) The role of ATP in mechanically stimulated rapid closure of the venuss-flytrap. Plant Physiol 51(1):17–18

Kaihara S, Takimoto A (1981) Effect of light and temperature on flower opening in *Pharbitis nil*. Plant Cell Physiol 22(2):215–221. https://doi.org/10.1093/oxfordjournals.pcp.a076158

Kier WR (2012) The diversity of hydrostatic skeletons. J Exp Biol 215(8):1247–1257. https://doi.org/10.1242/jeb.056549

Knight K (2014) Butterflies manipulate proboscis to suck. J Exp Biol 217(12):2031. https://doi.org/10.1242/jeb.108803

Krenn HW (1990) Functional morphology and movements of the proboscis of Lepidoptera (Insecta). Zoomorphology 110(2):105–114. https://doi.org/10.1007/BF01632816

Krenn HW, Mühlberger N (2002) Groundplan anatomy of the proboscis of butterflies (Papilionoidea, Lepidoptera). Zoologischer Anzeiger—A J Compar Zool 241(4):369–380. https://doi.org/10.1078/0044-5231-00078

Kornev KG, Monaenkova D, Yore C, Klipowics C, Edmond K, Sa V, Andrukh T (2009) Butterfly proboscis as a biomicrofluidic system. American Physical Society, 62nd Annual Meeting of the APS Division of Fluid Dynamics

Lehmann FO, Pick S (2007) The aerodynamic benefit of wing-wing interaction depends on stroke trajectory in flapping insect wings. J Exp Biol 210:1362–1377. https://doi.org/10.1242/jeb.02746

Lehnert MS, Monaenkova D, Andrukh T, Beard CE, Adler PH, Kornev KG (2013) Hydrophobic–hydrophilic dichotomy of the butterfly proboscis. J Roy Soc Interf 10(85):20130336. https://doi.org/10.1098/rsif.2013.0336

Lin HT, Slate DJ, Paetsch CR, Dorfmann AL, Trimmer BA (2011) Scaling of caterpillar body properties and its biomechanical implications for the use of a hydrostatic skeleton. J Exp Biol 214(7):1194–1204. https://doi.org/10.1242/jeb.051029

Lissmann HW (1950) Rectilinear locomotion in a snake (Boa *Occidentalis*). J Exp Biol 26(4):368–379

Llorens C, Argentina M, Rojas N, Westbrook J, Dumais J, Noblin X (2015) The fern cavitation catapult: mechanism and design principles. J Roy Soc Interf 13. https://doi.org/10.1098/rsif.2015.0930

Marmottant P, Ponomarenko A, Bienaimé D (2013) The walk and jump of Equisteum spores. In: Proceedings of the Royal Society B, 280, pp. 20131465. https://doi.org/10.1098/rspb.2013.1465

Marvi H, Hu DL (2012) Friction enhancement in concertina locomotion of snakes. J R Soc Interface 9:3067–3080. https://doi.org/10.1098/rsif.2012.0132

Marvi H, Bridges J, Hu DL (2013) Snakes mimic earthworms: propulsion using rectilinear travelling waves. J Roy Soc Interf 10(84). https://doi.org/10.1098/rsif.2013.0188

McCurley RS, Kier WM (1995) The functional morphology of starfish tube feet: the role of a crossed-fiber helical array in movement. Biol Bull 188(2):197–209. https://doi.org/10.2307/1542085

Mezoff S, Papastathis N, Takesian A, Trimmer BA (2004) The biomechanical and neural control of hydrostatic limb movements in Manduca sexta. J Exp Biol 207(17):3043–3053. https://doi.org/10.1242/jeb.01136

Monaenkova D, Lehnert MS, Andrukh T, Beard CE, Rubin B, Tokarev A, Lee WK, Adler PH, Kornev KG (2011) Butterfly proboscis: combining a drinking straw with a nanosponge facilitated diversification of feeding habits. J Roy Soc Interf 9(69). https://doi.org/10.10898/rsif.2011.0392

Moon B (2010) Snake locomotion. University of Louisiana at Lafayette. Accessed 25 April 2016 from http://www.ucs.louisiana.edu/~brm2286/locomotn.htm

Nachtigall W (2005) Biologisches Design, Systematischer Katalog für bionisches Gestalten. Springer Verlag, Berlin. ISBN 978-3-540-27369-1

Newcombe F (1888) Spore-dissemination of equisetum. Bot Gaz 13(7):173–178. https://doi.org/10.1086/326297

Noblin X, Rojas NO, Westbrook J, Llorens C, Argentina M, Dumais J (2012) The fern sporangium: a unique catapult. Science 335(6074):1322. https://doi.org/10.1126/science.1215985

Pandolfi C, Masi E, Voigt B, Mugnai S, Volkmann D, Mancuso S (2014) Gravity affects the closure of the traps in *Dionaea muscipula*. BioMed Research International 2014, Hindawi Publishing Corporation. https://doi.org/10.1155/2014/964203

Phillips HL, Kende H (1980) Structural changes in flowers of *Ipomoea tricolor* during flower opening and closing. Protoplasma 102(3–4):199-215. Springer Verlag, https://doi.org/10.1007/bf01279588

Pritchard J (2001) Turgor pressure. Encyclopedia of life sciences. Nature Publishing Group, DOI. https://doi.org/10.1038/npg.els.0001687

Pritts MB, Rahn CD (2004) Design of an artificial muscle continuum robot. In: Proceedings of the IEEE international conference of robotics and automation, Piscataway, NJ, vol 5, pp 4742–4746. https://doi.org/10.1109/robot.2004.1302467

Quillin KJ (1999) Kinematic scaling of locomotion by hydrostatic animals: ontogeny of peristaltic crawling by the earthworm *Lumbricus terrestris*. J Exp Biol 202(6):661–674

Quillin KJ (2000) Ontogenetic scaling of burrowing forces in the earthworm *Lumbricus terrestris*. J Exp Biol 203(18):2757–2770

Ribak G, Weihs D (2011) Jumping without using legs: the jump of the click-beetles (*Elateridae*) Is morphologically constrained. PLoS ONE 6(6):e20871. https://doi.org/10.1371/journal.pone.0020871

Ritter A (2007) Smart materials in architecture, interior architecture and design. Birkhäuser, Basel. ISBN 13: 978-3-7643-7327-6

Rost TL (1997). Rice anatomy. Virtual Crops Anatomy Atlas, UC Davies University of California, Department of Plant Biology. Accessed on 26 March 2016 from http://www-plb.ucdavis.edu/labs/rost/rice/ricehome.html

Routier-Kierzkowska AL, Weber A, Kochova P, Felekis D, Nelson BJ, Kuhlemeier C, Smith RS (2012) Cellular force microscopy for in vivo measurements of plant tissue mechanics. Plant Physiol 158:1514–1522. https://doi.org/10.1104/pp.111.191460

Sane SP (2003) The aerodynamics of insect flight. J Exp Biol 206(23):4191–4208. https://doi.org/10.1242/jeb.00663

Schleicher S, Lienhard J, Poppinga S, Speck T, Knippers J (2010) Abstraction of bio-inspired curved-line folding patterns for elastic foils and membranes in architecture. Design and nature V, WIT Press, pp 479–489. https://doi.org/10.2495/dn100431

Seymour MK (1969) Locomotion and coelomic pressure in *Lumbricus terrestris* L. J Exp Biol 51 (1):47–58

Tremblay I, Guderley HE, Himmelman JH (2012) Swimming away or clamming up: the use of phasic and tonic adductor muscles during escape responses varies with shell morphology in scallops. J Exp Biol 215(23):4131–4143. https://doi.org/10.1242/jeb.075986

Trueman ER (1967) The dynamics of burrowing in Ensis (Bivalvia). Proc Roy Soc London B 166 (1005):459–476. https://doi.org/10.1098/rspb.1967.0007

Tsai CC, Monaenkova D, Beard CE, Adler PH, Kornev KG (2014) Paradox of the drinking-straw model of the butterfly proboscis. J Exp Biolo 217:2130–2138. https://doi.org/10.1242/jeb.097998

Uehara K, Kurita S (1989) An ultrastructural study of spore wall morphogenesis in equisetum arvense. Am J Bot 76(7):939–951. https://doi.org/10.2307/2444515

van Doorn WG, van Meeteren U (2003) Flower opening and closure: a review. J Exp Bot 54 (389):1801–1812. https://doi.org/10.1093/jxb/erg213

van Griethuijsen LI, Trimmer BA (2009) Kinematics of horizontal and vertical caterpillar crawling. J Exp Biol 212(10):1455–1462. https://doi.org/10.1242/jeb.025783

Volkov AG, Adesina T, Markin VS, Jovanov E (2008) Kinetics and mechanism of *Dionaea muscipula* trap closing. Plant Physiol 146(2):694–702. https://doi.org/10.1104/pp.107.108241

Volkov AG, Foster JC, Baker KD, Markin VS (2010) Mechanical and electrical anisotropy in Mimosa pudica pulvini, Landes Bioscience. Plant Signal Behav 5(10):1211–1221. https://doi.org/10.4161/psb.5.10.12658

Xiang JJ, Zhang GH, Qian Q, Xue HW (2012) Semi-rolled leaf 1 Encodes a Putative Glycosylphosphatidylinositol-Anchored protein and modulates rice leaf rolling by regulating the formation of bulliform cells. Plant Physiol 159:1488–1500. https://doi.org/10.1104/pp.112.199968

Yang R, Lenaghan SC, Zhang M, Xia L (2010) A mathematical model on the closing and opening mechanism for Venus flytrap, Landes Bioscience. Plant Signal Behav 5(8):968–978. https://doi.org/10.4161/psb.5.8.12136

Young J, Walker SM, Bomphrey RJ, Taylor GK, Thomas ALR (2009) Details of insect wing design and deformation enhance aerodynamic function and flight efficiency. Science 325 (5947):1549–1552. https://doi.org/10.1126/science.1175928

Zou L, Sun X, Zhang Z, Liu P, Wu J, Tian C, Qiu J, Lu T (2011) Leaf rolling controlled by the homeodomain leucine zipper class IV gene *Roc5* in rice. Plant Physiol 156:1589–1602

References

The manufacturer's authorised representative in the EU is Springer Nature Customer Service Centre GmbH, Europaplatz 3, 69115 Heidelberg, Germany. If you have any concerns regarding our products, please contact ProductSafety@springernature.com

Printed and bound by CPI Group (UK) Ltd, Croydon, CR0 4YY
22/09/2026
02221064-0006